Life in inland waters

W. D. Williams
PhD, DSc, DipEd
Professor of Zoology
University of Adelaide, South Australia

Blackwell Scientific Publications
MELBOURNE OXFORD LONDON EDINBURGH BOSTON

© 1983 by
Blackwell Scientific Publications
Editorial offices:
99 Barry Street, Carlton
   Victoria 3053, Australia
Osney Mead, Oxford OX2 OEL
8 John Street, London WC1N 2ES
9 Forrest Road, Edinburgh EH1 2QH
52 Beacon Street, Boston
   Massachusetts 02108, USA

DISTRIBUTORS

USA
   Blackwell Mosby Book Distributors
   11830 Westline Industrial Drive
   St Louis, Missouri 63141

Canada
   Blackwell Mosby Book Distributors
   120 Melford Drive, Scarborough
   Ontario M1B 2X4

Australia
   Blackwell Scientific Book Distributors
   31 Advantage Road, Highett
   Victoria 3190

Cataloguing in Publication Data

Williams, W. D. (William David), 1936-
   Life in inland waters.

   Bibliography.
   Includes index.
   ISBN 0 86793 088 8.

   1. Freshwater biology — Australia.
   2. Limnology — Australia. I. Title.

574.92′9′94

First published 1983

Printed by Brown Prior Anderson Pty Ltd Burwood Victoria

Contents

Preface

There are several books available that deal with the general ecological features of inland waters, and indeed there is one that treats that subject from an Australian perspective. There are also several books that describe particular elements of the Australian aquatic fauna and flora (e.g. macrophytes, invertebrates, fish, amphibians). There is not, however, any account that integrates these separate descriptions nor comprehensively describes Australian inland waters and their life from the viewpoint of a biologist. This book is an attempt to do so.

In writing it, I have had much pleasure (as well as the usual share of anxieties involved in such endeavours). A special source of pleasure has been the thought that I have been, in a sense, partly discharging a debt for a rich and rewarding professional life. I have worked for the past 20 years in a vast Aladdin's cave in the company of many robust and tolerant friends. It has been a wonderful experience. I came to Australia in 1961, when there were few Australian freshwater biologists, and rather little was known about Australian inland waters. Twenty years later, the number of freshwater biologists runs to some hundreds and Australian limnology is a vigorous, expanding and sophisticated scientific discipline. It has been a rare privilege to have become an Australian during such exciting times.

An additional pleasure has been the splendid cooperation received from many colleagues on whom I have inflicted various drafts. Mere thanks in a paragraph acknowledging their assistance seems little reward for the many hours spent helping me. Special thanks go to Dr K. F. Walker of the University of Adelaide who has read most chapters, and for whose incisive criticism and level-headed judgements I have over the years or our association developed a very healthy respect. I am indebted also to Dr M. C. Geddes, also of the University of Adelaide, for comments on a final overview of the whole book. My thanks, likewise, go to the numerous colleagues who have read individual chapters from the standpoint of experts in their field. I thank in this regard Miss H. I. Aston, Royal Botanic Gardens and National Herbarium, Melbourne (macrophytes), Dr Shelley Barker, University of Adelaide (mammals), Dr J. Bauld, Bass-Becking Geobiological Laboratory, Canberra (microscopic plants), Dr L. W. Braithwaite, CSIRO Division of Wildlife Research, Canberra (birds), Dr J. Bowler, Australian National University, Canberra (palaeolimnology), Mr I. Campbell, Chisholm Institute of Technology, Melbourne (conservation), Dr P. De Deckker, Australian National University, Canberra (invertebrates, palaeolimnology), Dr G. G. Ganf, University of Adelaide (microscopic plants), Dr H. B. N. Hynes, University of Waterloo, Canada (conservation), Dr P. S. Lake, Monash University, Melbourne (conservation), Dr D. S. Mitchell, CSIRO Division of Irrigation Research (macrophytes), Mr S. Parker, South Australian Museum, Adelaide (birds), Mr L. F. Reynolds, New South Wales State Fisheries, Sydney (fish), Dr A. C. Robinson, National Parks and Wildlife, Adelaide (reptiles), Dr R. S. Seymour, University of Adelaide (reptiles), Dr. R. J. Shiel, University of Adelaide (rivers), Dr Gurdip Singh, Australian National University, Canberra (palaeolimnology), Dr D. R. Towns, University of Adelaide (rivers and streams), Mr M. J. Tyler, University of Adelaide (amphibians), and Dr P. A. Tyler, University of Tasmania (microscopic plants). I have nearly always accepted the advice of these colleagues, but not invariably. Errors of omission and fact are therefore squarely mine.

I also acknowledge help from many colleagues and institutions who have provided photographs for reproduction, unpublished information, comments on sections of individual chapters or otherwise supported my efforts. I thank in this respect Australian Information Service (Canberra), Dr T. M. Berra (Ohio State University, USA), Dr I. Beveridge (Institute of Medical and Veterinary Science, Adelaide), Dr B. C. Chessman (Latrobe Valley Water and Sewerage Board, Taralgon), Miss M. Davies (University of Adelaide), Dr D. Griffin (Australian Museum, Sydney), Dr G. Grigg (University of Sydney), The Institute of Medical and Veterinary Science (Adelaide), Dr P. Jackson (Fisheries and Wildlife Division, Victoria), Dr J. C. Jessop (Botanic Gardens and State Herbarium, Adelaide), Mr T. Lowe (Mystic Park, *via* Kerang), Mr B. J. McMahon (Queensland Institute of Technology, Brisbane), Mr G. Miles (Northern Territory Administration, Darwin), Dr A. C. Robinson (National Parks and Wildlife, Adelaide), Dr P. J. Stanbury (University of Sydney), Dr P. D. Temple-Smith (Australian National University), Dr K. F. Walker (University of Adelaide), and Mr J. C. F. Wharton (Fisheries and Wildlife Division, Victoria). The cover illustration was provided by the Tasmanian Wilderness Society to which I extend my thanks.

Special acknowledgement is made of the support of three more of my colleagues at the University of Adelaide. I thank Miss S. Lawson, Departmental Secretary, for translation of my scruffy drafts and near illegible handwriting to impeccably typed manuscripts. I thank Mr Phil Kempster for the production of photographs from a motley collection of slides, negatives and old prints.

And I thank Ms Ruth Altmann for turning my rough sketches into respectable figures.

A note should be added concerning references. To make reading easier, references have not been included in the text. This perhaps creates an impression of originality. May I dispel that impression. The book is founded on the work and ideas of others and on some previously published work of my own; very little of it is original in the sense that ideas are put forward which are not derived from published work. A list of the books and papers on which each chapter is based is given at the back of the book.

I have tried to produce a text that will be understood by non-specialists. In doing so, I have therefore constantly tried to bear in mind the advice of H. G. Wells to Julian Huxley:

> The reader for whom you write
> is just as intelligent as you are but
> does not possess *your* store of knowledge,
> He is not to be offended by a recital
> in technical language of things known to him . . .
> He is not a student preparing for
> an examination and *he does not want to be*
> *encumbered with technical terms*,
> his sense of literary form and his sense of humour is probably
> greater than yours.

W. D. Williams   *Adelaide, 1982*

1 Introduction

Australia is arid. After Antarctica, it is the driest continent, with an average annual rainfall only two-thirds that for all land areas of the world, and an average annual run-off about one-eighth the world average. This does not mean that Australia lacks surface waters. It does mean that they are not evenly distributed, that total volumes are low, and that careful husbandry of water resources is required.

Yet, what Australia's surface waters lack in quantity is more than compensated for in diversity, scientific interest and beauty. Fresh waters and saline waters, small ephemeral pools and deep permanent lakes, fast flowing upland streams and languid lowland rivers: all occur. Note also that wet tropical regions, hot deserts, temperate lowlands, semi-arid plains, and cold highlands are major Australian environments. This physical diversity combines with the long geographical isolation of the continent, its former connections with an ancient southern landmass (Gondwanaland), and a proximity to southeast Asia to produce an aquatic fauna and flora that is at once both distinctive and of great scientific interest.

On the face of it, then, one might expect Australia's lakes, rivers and other surface waters to have been the subject of much concern and interest: to have been relatively well-explored by scientists interested in inland waters (limnologists), to be regarded as treasured items of the nation's natural heritage, and to have been managed wisely and prudently. Unfortunately, this is not so. As later chapters will reveal, we know remarkably little about the ecology of Australian inland waters, given the size and affluence of the community. As well, many water-bodies have been destroyed or polluted. And water management is beset by problems difficult to resolve within the present framework of governmental attitudes, science and administration.

The picture is not quite as bleak as the preceding paragraph might suggest, however. Numerous scientists are now studying Australian inland waters. Conservationists are actively trying to preserve them. And governments are recognizing that, as a basic resource in short supply, Australian inland waters need better management.

Scientific interest has grown mostly in the past two decades. A good barometer of it is the size of the membership of the Australian Society for Limnology, a professional body founded in 1962. From an initial membership of less than 50, the number of members has grown to almost 400. Scientific knowledge has grown correspondingly, though this reflects more the dedication of limnologists than any greatly increased financial support for limnology. Indeed, paradoxically in an arid continent, limnology is the poor sister of marine science.

Likewise, the last two decades have seen the growth of a greatly increased sensitivity and awareness on the part of the Australian community to matters of environmental conservation concerning natural waters. Indeed, the most vehement of all Australian conservation battles involved a lake: the fight by conservationists to prevent the flooding of Lake Pedder, Tasmania. The conservationists lost that battle, but there is no doubt that they had wide community support. Even now another battle is underway: the attempt to prevent the damming of the Franklin River in Tasmania. Again there is enormous support from the community for the conservationists, but the outcome as yet is unknown. Whatever the result, the attitudes of governments have changed in response to changed community values, and most now recognize their obligations to conservation.

Recognition by governments of the need for better management of water resources has been slower to materialize. There is no doubt, however, that it is there; the number of recent government enquiries into management problems is firm evidence of it. Of course, recognition of the need for better management is one thing; implementation of ideas to bring it about is quite another. So far, very few ideas have been implemented. A major difficulty is that management is in the hands of individual State governments, each operating with local, and sometimes conflicting objectives in mind. The problems associated with management of the River Murray are a case in point. A second major difficulty is that knowledge for wise management is inadequate. This particular difficulty may lessen shortly, for several proposals are being canvassed which may help provide such knowledge. They include proposals for an Institute of Freshwater Studies, for an expanded programme of research by CSIRO, and for the establishment of regional centres to investigate water resource problems. None of these proposals may, of course, be implemented — at least in the short term. Whatever the outcome, the demand for water is increasing, its availability is not. Increasing management problems will inevitably follow and force better management in the future, if not now. That much is certainly recognized by governments.

Dissatisfaction with the general state of affairs outlined in preceding paragraphs partly prompted this book. Of paramount importance, however, was a particular dissatisfaction with the lack of recognition accorded the scientific interest of Australian inland waters. Few writers have emphasized this interest, and, of those who have, most have relied more on sweeping generalizations or particular features to support their case than a comprehensive documentation of the evidence. This book attempts to rectify matters.

For the most part, the approach is unashamedly descriptive and parochial. Readers who wish to know about general limnological arrangements and processes must look elsewhere; there are several excellent texts available (see References). The major emphasis here is upon describing the special features of Australian inland waters and their life, and on describing the diversity of both.

The principal types of inland water are considered in chapters 2 (standing waters) and 3 (flowing waters). Particular elements of the fauna and flora are considered in chapters 4–10: invertebrates (4), fish (5), amphibians and reptiles (6), birds (7), mammals (8), microscopic plants and allied organisms (9), and macrophytes (10). The remaining chapters consider three more general topics that are of special significance with regard to overall aims. Chapter 11 is concerned with the subject of conservation. Chapter 12 outlines the nature of man's impact. And the final chapter looks at environments from an historical viewpoint.

Throughout the text a special effort has been made to draw attention to the many gaps in knowledge. The hope is that not only will readers be better informed about Australian inland waters after reading the book, but that they will also be stimulated to add to what we know about them.

2   Lakes and other standing waters

All inland surface waters are in motion, but a distinction can be drawn between those where currents are not easily discernible, standing waters, and those where they are, running waters. Familiar examples of standing waters are lakes, reservoirs, rain-pools and farm dams, and of running waters, rivers and streams. The present chapter considers standing waters; running waters are considered in the next chapter.

SPECIAL FEATURES

In general nature, Australian standing waters are like those elsewhere. However, they also have many special features, discussion of which follows.

Given the immense area of Australia, an obvious special feature is the paucity of standing waters: reflecting continental aridity, lakes and other natural standing waters are not numerous. There are a few, scattered mainland regions where permanent lakes are common (notably, the Western District of Victoria), but the large blue areas on many maps of Australia are quite misleading! They usually indicate areas where large bodies of standing water occasionally occur. It is in Tasmania alone that permanent bodies of standing water are widespread. Indeed, ephemerality is another common attribute of Australian standing waters.

Two other physical features are noteworthy: patterns of thermal behaviour within relatively deep water-bodies, and light penetration.

In all surface standing waters, whilst currents may not be easily discernible, water is nevertheless in motion. This motion is largely powered by wind, with downwards transmittal of energy by counter currents. Significant modifications of water movement may occur in all but the most shallow of water-bodies because water stratifies under the influence of temperature. When this happens, two layers develop: an upper, warmer and less dense layer, the epilimnion, and a lower, colder and denser layer, the hypolimnion. Water currents continue in the epilimnion, but are reduced in the hypolimnion which is more isolated from surface events. The pattern of stratification may vary daily, as in shallow water-bodies, or over longer periods, as in deep lakes. Longer term patterns are determined by many factors, important ones being lake depth, climate, and degree of exposure; the two commonest patterns in Australia are the so-called polymictic one and the warm-monomictic one. In polymictic lakes, stratification is not seasonally persistent, and may last just days in almost any season; in warm-monomictic lakes, stratification develops in summer, but in other seasons water circulates completely (Fig. 2.1). So far no published records exist of the dimictic pattern, involving two periods of complete circulation (spring and autumn). Dimictic lakes are widespread in temperate Europe and North America, and are the sort of lake most studied and upon which many limnological concepts are based. There are reports of dimictic lakes in Tasmania and in highland areas of the mainland, but even if confirmed, it is still clear that the general absence of dimictic lakes from temperate regions of Australia is a noteworthy feature.

A feature of many Australian lakes and reservoirs is that light does not penetrate far: they are highly turbid and coloured. The high turbidity has long been recognized by crude measurement, but the nature of light penetration into them is now being put on to a precise scientific footing. Typically, there are high concentrations of suspended particulate matter; this material scatters light of which much is absorbed by gilvin, the yellow dissolved organic substances which strongly colour the water. According to the amounts of the three light-absorbing components present (gilvin, inorganic particles and plankton), Australian standing waters have been tentatively assigned to three main types: those with plenty of gilvin, but of low turbidity (type 'G'), those with gilvin and high concentrations of plankton ('GA'), and those of high turbidity ('T'). This classification is, of course, neither static nor comprehensive.

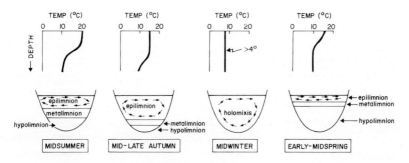

Fig. 2.1 Pattern of thermal stratification in many Australian lakes and reservoirs: the warm monomictic pattern. Redrawn and modified from Bayly and Williams (1973).

Quite why so many standing waters are turbid and coloured is uncertain. High turbidity is certainly a result of shallowness, and the ease with which the wind can resuspend sediment. The sediment, and probably also the gilvin, are derived from catchment soil erosion.

Two chemical features are of interest: salinity and chemical composition. High salinities characterize many Australian lakes, and salt lakes (roughly those with salinities greater than 3°/oo or 3 g/l) occur in all states. Natural bodies of *fresh* standing water of any permanency are common only in the higher rainfall areas of northern Australia, the east coast, the temperate southeast (including Tasmania) and the extreme southwest.

The saline waters are remarkably homogeneous in chemical composition; the dominant elements are sodium and chloride (in a few cases, (bi)carbonate ions may be important). Since this composition is like that of sea water, it has led to suggestions that much of the salt has come from the sea by 'cyclic accession', that is, by inland transport of sea spray salt in rain and dust. For many salt lakes, 'cyclic salt' is undoubtedly an important source, but former marine sediments and saline groundwater must also be significant in certain areas. Whatever their origin, sodium and chloride are clearly important elements in Australian inland waters. This is so for many fresh waters too, since their chemical composition frequently departs from the so-called standard composition of fresh water, in which magnesium, calcium and (bi)carbonate ions dominate, towards a composition dominated by sodium and chloride. Table 2.1 provides analyses to illustrate these points.

The many special biological features of Australian standing waters are detailed in later chapters. The fauna in particular is characterized by high levels of endemicity, the absence of certain groups common elsewhere, the presence of groups unknown or not abundant elsewhere, and great diversity in yet other groups.

Table 2.1 Chemical composition of some Australian lakes. Except salinity, data expressed as percentages of total cation or anion sum in equivalents. Analyses have been selected over a range of salinities (fresh to salt lakes) to illustrate variety of ionic dominances. Dominant ions for each lake are in bold. Data derived from **several** sources.

Lake	Salinity	Ions						
	°/oo	Na	K	Ca	Mg	Cl	SO_4	$HCO_3 + CO_3$
Weering (Vic.)	213	**85**	1	< 1	14	**96**	4	< 1
Grace (WA)	60	**86**	1	2	11	**94**	5	< 1
Werowrap (Vic.)	39	**95**	4	0	1	**66**	0	34
Leake (SA)	2	**74**	2	4	20	**75**	8	17
Barrine (Qld)	0.05	30	5	19	**46**	35	0	**65**
Woods (NT)	0.05	10	11	4	**75**	7	19	**74**
Cootapatamba	0.003	**70**	8	12	10	**37**	30	33

PRINCIPAL TYPES OF STANDING WATER

No firm boundaries exist between the various sorts of standing water, but a limited number of principal types may be distinguished: salt lakes, freshwater lakes and reservoirs, billabongs, farm dams, wetlands, and rain-pools and puddles. In addition, another type of more heterogeneous composition is distinguished for certain other small bodies of fresh water. The following account is uneven because we know more about some types than others.

Salt lakes

Salt lakes occur in all Australian states, but are particularly common in certain regions. They are numerous in southern Western Australia in an area named 'Salinaland' by geomorphologists; nearly all are temporary, shallow localities. As a region of salt lakes, Salinaland may be as old as 100 million years. Numerous salt lakes also occur in western Victoria. Here they range from shallow, temporary localities to deep, permanent lakes. Many of the larger ones result from volcanic activity during the past 50 000 years. In South Australia, there are many near the coast as well as many large ones in central arid regions. The coastal ones are mostly shallow, originate from relatively recent geological uplifting, and are seasonally replenished with water. Those inland contain water much less regularly and stem from older geological events.

Fig. 2.2 Salt lakes. (a) Lake Corangamite, Victoria; (b) Lake Eyre, South Australia; (c) Lake Bullenmerri, Victoria; (d) Lake Cowan, Western Australia.

Table 2.2 Dimensions of some notable Australian lakes. Data from various sources.

Lake	Type	Surface area (ha)	Volume (m³ × 10⁶)	Maximum depth (m)	Mean depth (m)
Corangamite (Vic.)	Salt	23 300	456	4.9	2.9
Bullenmerri (Vic.)	Salt	488	192	66	39.3
Gnotuk (Vic.)	Salt	208	32	18.5	15.3
Great Lake (Tas.)*	Fresh	16 000	2 400	20	15
St Clair (Tas.)*	Fresh	2 800	2 000	165	71
Eyre (SA)†	Salt	860 000	26 000	—	3

* Natural levels increased for hydroelectric purposes.
† Mostly dry (all other lakes are permanent).

Figure 2.2 illustrates a little of the physical diversity. Lake Corangamite is the largest permanent lake in Australia, and Lake Bullenmerri amongst the deepest. Lake Eyre fills irregularly, but may spread its waters over a vast area. Information on the dimensions of these and other notable Australian lakes is given in Table 2.2.

Salinities range from a lower arbitrary limit of 3°/∘∘ to about 350°/∘∘, ten times sea water salinity. Although large permanent lakes, such as Lake Gnotuk (Fig. 11.4(b); about 70°/∘∘), have more or less constant salinities over a year, wide seasonal fluctuations usually occur, with highest salt concentra-

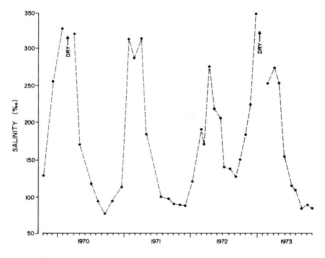

Fig. 2.3 Seasonal fluctuations in salinity in a temporary saline lake in Victoria. Redrawn after Williams and Buckney (1976).

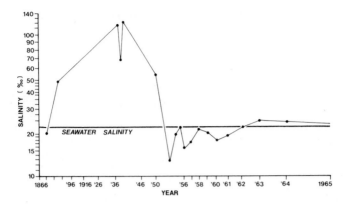

Fig. 2.4 Long-term variations in the salinity of Lake Corangamite, Victoria. Redrawn after Bayly and Williams (1966).

tions — often, saturation levels — in summer, and low ones in late winter (Fig. 2.3). There may also be longer term trends, best illustrated by Lake Corangamite (Fig. 2.4). The dominant elements are sodium and chloride, as mentioned (Table 2.1).

The fauna comprises three major groups: 'halobionts' characteristic of highly saline waters (salinities in excess of $50°/oo$), 'halophiles' in moderately saline waters ($10–60°/oo$), and 'salt-tolerant' freshwater animals (below $20°/oo$). As a *general* rule, the higher the salinity, the fewer the species present. Important halobionts are *Haloniscus searlei* (an isopod; Fig. 4.3(d)), larvae of *Tanytarsus barbitarsis* (a chironomid), and several species of ostracods, calanoid copepods and *Parartemia* (brine shrimp). Important halophiles are *Daphniopsis pusilla* (a water-flea), *Coxiella* (a snail), rotifers, ostracods and some insect species. Much greater numbers of species are found in the third group, and include many insect species.

Three important points should be made about the faunal composition. First, it is different from that elsewhere; *Parartemia, Coxiella, Haloniscus* and several other forms are endemic to Australian salt lakes. *Artemia 'salina'* and ephydrid fly larvae, frequently regarded as typical of salt lakes worldwide, are not common (*A. 'salina'* is probably an introduction to coastal salt pans). Second, marked regional differences occur; in Western Australia, for example, there is a number of species of *Coxiella* and *Parartemia*, whereas in Victoria these genera have a single species, not present in the west. And third, the evolutionary relationships of the fauna are predominantly with the fauna of fresh waters. To underscore this, the term 'athalassic' was coined to refer, broadly, to inland waters with no direct marine connection. The term is particularly useful when referring to coastal salt lakes.

The fauna displays many adaptations to cope with the stresses of life in salt lakes (most notably high salinities and frequent desiccation). Adaptations to high salinity primarily involve the ability to tolerate high body-fluid concentrations (as in osmoconformers) or the ability to regulate body-fluid concentrations (osmoregulators). Adaptations to desiccation mainly involve the production of resistant eggs, though in a few species adults survive drying.

The flora is characterized by a paucity of species. In highly saline lakes, macrophytes are absent, and the phytoplankton is dominated by a green alga, *Dunaliella salina*. In less saline waters, the phytoplankton includes a variety of diatoms, green algae other than *Dunaliella*, and several Cyanobacteria (especially species of *Anabaena, Microcystis, Spirulina* and *Nodularia*). A few species of macrophyte — *Ruppia* (Fig. 10.3(d)), a grass-like plant, is the most important — may also be present (see chapter 10). Rates of phytoplankton production span a considerable range: values of 24–2201 mg carbon fixed/m²/year have been recorded. How most salt lake plants cope with high salinity is unknown. However, *Dunaliella salina* does it by increasing internal cellular concentrations with glycerol (chapter 9), and *Ruppia* by using proline (an amino acid) in the same way.

Salt lakes, in Australia as elsewhere, are little studied, given their abundance and distribution. This situation is changing with increasing recognition of three facts. Their simplified biological composition provides unique ecological opportunities for studies of whole ecosystems. The adaptations of their biota provide unique oportunities for study by physiologists. And their sensitivity to climatic change provides unique palaeolimnological opportunities. Here, then, is a special opening for Australians to contribute significantly to knowledge of inland waters.

Freshwater lakes and reservoirs

There are some large, permanent freshwater lakes in Australia (Table 2.2, Fig. 2.5), but, except in Tasmania, they are not usual features of the landscape. A few are found in the extreme southwest of Western Australia, in South Australia near Mount Gambier, on the volcanic plains of western Victoria, and near the south coast (Gippsland Lakes) of eastern Victoria. New South Wales has some shallow and widely dispersed lakes east of the Dividing Range (notably Lakes George and Cowal), but, otherwise, freshwater lakes are mainly associated with siliceous dunes of the northeastern coast or confined to highlands near Mount Kosciusko. In Queensland, likewise, there are a few shallow and widely dispersed lakes east of the Dividing Range (e.g. Lake Dunn), but most freshwater lakes are generally not far from the east coast; two important groups are those on the Atherton Tablelands in the far northeast (e.g. Lake Euramoo, Figs. 10.7, 11.4(c)), and those associated with

Fig. 2.5 Freshwater lakes. (a) Blue Lake, South Australia; (b) Lake Dunn, Queensland; (c) Lake Lilla, Tasmania; (d) Lake Dobson, Tasmania.

coastal sand dunes (Fig. 11.3(b)). Finally, in the Northern Territory, the only sizeable freshwater lake of any permanency is Lake Woods.

This natural distribution has been much altered by the construction of impoundments (reservoirs) to serve agricultural, industrial and domestic needs. The effect has been not only to increase the number of deep, large freshwater bodies, but to extend their distribution. Thus, there are now several large reservoirs in southwestern Western Australia, as well as a huge one (Lake Argyle) in the north of that state. Barrages across the mouth of the River Murray in South Australia have converted Lake Alexandrina from a brackish coastal lake to a freshwater impoundment. And reservoirs now exist in highland regions of Victoria, New South Wales (Fig. 12.6) and Queeensland, as well as in areas inland of the Dividing Range. Even in Tasmania, many reservoirs have been created or lake levels artificially elevated. So far, over 300 large impoundments have been constructed throughout Australia. The ten largest are listed in Table 2.3.

Australian freshwater lakes have had many origins. Pleistocene glacial action gave rise to most of those in the high country of Tasmania and to the five near the summit of Mount Kosciusko on the mainland (otherwise, the recent Ice Age was not a major limnological event in Australia, as in northern Europe and North America). Large scale earth movements (tectonic movements) probably gave rise to Lake George in New South Wales and the Great Lake of Tasmania. Smaller crustal movements, a gradual sinking of the

Table 2.3 Capacity and height of the 10 largest Australian reservoirs. Natural lakes now further impounded are excluded. Cautionary note: reservoirs are now officially referred to as Lake _____. Data from various sources.

Reservoir	Capacity ($m^3 \times 10^6$)	Height of dam (m)
Gordon (Tas.)	11 700	140
Argyle (WA)	5 700	67
Eucumbene (NSW)	4 800	116
Dartmouth (Vic.)	3 700	180
Eildon (Vic.)	3 400	79
Hume (NSW)	3 100	43
Serpentine (Tas.)	2 900	42; 43
Burragorang (NSW)*	2 100	115
Burrendong (NSW)	1 700	76
Blowering (NSW)	1 600	112

* = Warragamba Dam

coastal plain, gave rise to the Gippsland Lakes. Volcanic activity was responsible for several lakes in western Victoria (e.g. Lake Purrumbete), those near Mount Gambier, and those on the Atherton Tablelands. Lake Tarli Karng in Victoria and a few small lakes elsewhere resulted from river blockage by landslides. Wind action, together with other events, seems to have been important in the formation of many coastal dune lakes and lakes in semi-arid inland regions. Finally, solution of rock strata has given rise to a few lakes, and fluviatile action to others (but see section on billabongs, below).

Little correlation generally exists between mode of origin and chemical nature. Nevertheless, the freshest lakes are those of glacial origin or on siliceous sand dunes. The lowest salinity recorded for any Australian lake is 3 mg/l, a value very near that of distilled water! It was recorded from the glacial lakes near Mount Kosciusko. Unlike many saline lakes (Fig. 2.3), the salinity of freshwater lakes does not undergo significant seasonal fluctuation.

Because there are many more plants and animals in freshwater lakes and reservoirs than in salt lakes, detailed consideration of them is left to later chapters. Only a few comments on two of the major animal communities (Fig. 2.6) are apposite.

The *benthos* (bottom living community), as in all freshwater lakes, decreases in diversity (and biomass) with depth, but overall its diversity is low compared with that in freshwater lakes elsewhere in the world. As a rule, it contains fewer than 50 species, whereas lakes in the northern hemisphere have more than 50. The relative importance of major groups is indicated in Table 2.4. The reasons for the low diversity are uncertain: geographical isolation, the

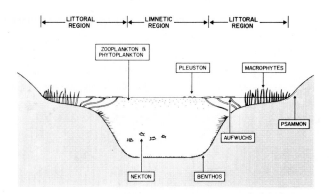

Fig. 2.6 Major biological communities of standing freshwater-bodies. The phytoplankton and zooplankton are microscopic, free-floating organisms; the nekton comprises large, free-swimming animals; the benthos comprises bottom-living animals; microscopic plants and animals living on or closely associated with submerged surfaces constitute the Aufwuchs; the surface film community is the pleuston; and the community inhabiting the interstices of littoral sediments is the psammon.

'youth' of many Australian lakes, their general paucity, and the lack of well-defined seasonal climatic differences have all been advanced as possible explanations. There are further points of interest; for instance, it is difficult to apply trophic schemes erected for lakes in the northern hemisphere and involving biomass:depth profiles, dominant chironomids, and relative contributions of different animal groups to total biomass.

Table 2.4 Percentage contribution of seven major groups of invertebrates to total biomass of the benthos in two series of freshwater lakes in southeastern Australia. Data derived from Timms (1980).

Major groups	Mean for 7 Tasmanian lakes	Mean for 4 Kosciusko lakes
Turbellaria	1	3
Oligochaeta	31	16
Crustacea	18	47
Chironomidae	41	32
Chaoborinae	1	0
Other insects	5	0
Mollusca	2	6

There is some evidence that the number of species in the zooplankton community of Australian freshwater lakes is likewise depressed: less than three calanoid copepod, two cyclopoid copepod, three cladoceran and four or five rotifer species usually occur. A more certain distinctive feature concerns the calanoids. Over the southern half of Australia these belong to the Centropagidae, a diverse and widespread family in fresh waters of the southern hemisphere, but one less well-developed elsewhere. The majority of Australian species is endemic. Increasingly greater endemicity is now being found in several zooplanktonic groups (rotifers, cyclopoids) formerly regarded as comprising, in the main, cosmopolitan species.

Of course, the zooplankton has many other features of interest in addition to its regional distinctiveness. For example, the recent discovery that notonectid bugs secrete a substance which can induce different morphological forms of some water-fleas (*Daphnia 'carinata'*) is of interest in attempts to explain so-called cyclomorphosis, cyclical changes in morphology of apparently seasonal occurrence. Cyclomorphosis is a feature of all zooplanktonic communities and occurs in many groups (rotifers, copepods) besides cladocerans. The investigation of the influence of notonectids grew from knowledge of another interesting phenomenon recently found in certain populations of Cladocera in Australian fresh waters: modern biochemical techniques (gel electrophoresis) could distinguish several 'species' with little if any morphological difference between them.

Billabongs

In several regions of Australia, small permanent water-bodies containing standing fresh water are closely associated with rivers. Some seasonally connect with their parent river, others less regularly or not at all. The colloquial name for such bodies is billabong, a word of Aboriginal derivation. The term is not used outside Australia, but is an apt, euphonious and comprehensive one of considerable utility.

Billabongs clearly link standing and flowing waters. Those regularly in connection with rivers stand at one end of the link, those irregularly in connection or now isolated stand at the other: on the one hand, billabongs merge into the category of large river pools, on the other, that of small lakes of fluviatile origin. Here, those which are seasonally and regularly replenished by rivers will be considered as flowing waters, and those not, as standing waters. Thus, billabongs of northern monsoonal regions are considered in chapter 3, whilst those of other regions are considered here. It need scarcely be added that this treatment is as much a matter of convenience as any recognition of exclusive differences between the two sorts of billabong.

Most Australian rivers which meander develop billabongs. These are particularly common in the southern part of the Murray–Darling system (Fig. 11.3(a)). Most are 'ox-bow lakes', formed by lateral displacement of sediment, differential erosion, and downstream migration of river 'loops'. Typically, a billabong of this system is shallow, crescentic and narrow; it is usually less than 5 m deep, up to several kilometres long, and not more than 100 m wide. With siltation and the gradual accumulation of plant material it eventually becomes a swamp and then dry land.

Physicochemical features vary enormously between billabongs, a major determinant being the pattern of water replenishment. As a general rule, however, turbidities are low (except at times of flood), water temperatures follow air temperatures with thermal stratification likely only in deep, sheltered billabongs, and salinities are relatively low though fluctuate according to flooding and rainfall patterns.

The biological communities are complex. The phytoplankton contains many species, and a characteristic feature is the development of a diverse assemblage of macrophytes in all lifeforms (see chapter 10). Moreover, the seasonal succession of macrophytes, their dominant species, and their overall species composition often differs in adjacent billabongs. The zooplankton invariably includes species more typical of littoral (marginal) aquatic habitats than of lakes; it usually comprises two or three species of calanoid copepod, two or more cyclopoid copepods, two or three Cladocera, and one to four species of rotifer. The littoral invertebrate fauna may be extremely diverse, particularly in crustaceans and rotifers. In one billabong near Alexandra, Victoria, for example, 85 species of these groups have been recorded, the greatest microfaunal diversity recorded from any freshwater habitat. Vertebrate members of the fauna are discussed in later chapters.

Finally, any consideration of billabongs should mention their important ecological role. They provide refuges, food, and breeding sites for a wide selection of invertebrates and vertebrates over large areas of arid inland Australia. Their security, from a conservation viewpoint, is a matter of great concern (see chapters 11 and 12).

Farm dams

Small bodies of standing water constructed for stock are found throughout Australia. Generally referred to as farm dams, or sometimes farm tanks, more than 400 000 may have been built. Many occur in areas formerly lacking permanent standing water (Fig. 12.5). Despite their ubiquity, large number, and the extension they represent in the geographical spread of standing waters, surprisingly little is known about them.

Four types can be recognized: gully dams, hillside dams, excavated tanks and raised-edge dams. Stored volumes may fluctuate widely, reflecting patterns of usage and seasonal differences in rainfall and evaporation. As a result, distinct seasonal fluctuations in salinity may also occur, although upper salinity values rarely exceed 3°/oo. Plant nutrient concentrations (phosphates, nitrates) are often high. Turbidity is also usually high. Water temperatures generally follow air temperatures, with thermal stratification rarely persistent (at least in small dams).

Zooplankton communities are similar to those of lakes and reservoirs. Benthic diversity and biomass are typically low, with chironomids, chaoborids and oligochaetes the dominant forms. Wherever stock have access to the water's edge, or when water-levels fluctuate greatly, littoral macrophytes are absent, with correspondingly low diversities of associated invertebrates. Much higher diversities of littoral invertebrates occur when macrophytes develop. Fish may be present as the result of stocking.

Waste stabilization ponds, or sewage lagoons, are a special type of farm dam. They are used in the treatment of wastewater, and improve quality by degrading complex organic material and decreasing pathogenicity. Their function is discussed further in chapter 9. Because of simple yet effective operation, municipal authorities thoughout Australia are making increasing use of them, and they now occur on the outskirts of many towns and other communities. Generally out of bounds to all but authorized personnel, they are of little interest in a book of this sort. However, in ponds that are not so heavily loaded as to become depleted of oxygen, phytoplankton diversity is reduced and macrophytes are generally absent. Chironomids are important members of the invertebrate community.

Wetlands

In certain lowland areas of Australia, large but shallow sheets of standing water exist for part of a year (Figs 2.7 (a), 11.2(b), (c)). Such ephemeral waterbodies are not necessarily associated closely with river systems. Some can perhaps best be regarded as seasonal swamps, and others as ephemeral lakes. Whatever the case, they possess a number of common characteristics sufficiently distinctive to merit a separate category of standing waters. They are aptly described as 'wetlands'.

Wetlands, like salt lakes, have been much less studied by limnologists throughout the world than their abundance and widespread distribution might suggest. They have been grossly neglected as study sites in Australia. Unfortunately, so many are being drained or otherwise destroyed — despite awareness of their ecological and hydrological importance — that in many regions it will soon be impossible to study them.

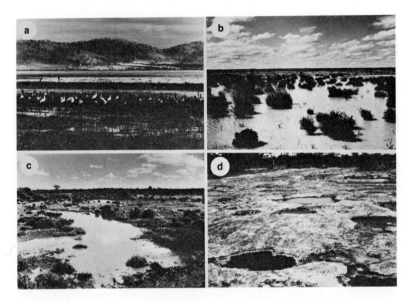

Fig. 2.7 Small bodies of standing fresh water. (a) Townsville Common (wetland), Queensland; (b) rain-pool near Lake Eyre, South Australia; (c) rain-pool (claypan) near Wiluna, Western Australia; (c) small granite rock-pools, near Yowerda Soak, Western Australia.

Wetlands span all parts of the hydrological spectrum: in some, water is present for just a short period, in others for most of the year. Over most of southern Australia, water is typically present during winter and spring, dries in summer, and reappears in autumn. Individual patterns are determined by geographical position, local hydrology and topography, and by climate. Light generally penetrates to the bottom, water temperatures follow air temperatures, and salinities gradually rise as water levels fall (but never far above 3°/∘∘).

Complex and diverse macrophyte communities are present, and generally extend across the entire width of localities. Zonation, however, may often be indistinct (see Fig. 10.10). Seasonal succession of macrophytes is an obvious feature in many wetlands, with truly aquatic species being succeeded by semi-aquatic forms, then, during the dry period, by shortlived terrestrial species. Additionally, some species are perennial, but these are mainly marginal in position.

The zooplankton and phytoplankton are diverse, but the benthic community lacks several prominent invertebrate groups, probably because they cannot tolerate periods of desiccation and have poor powers of dispersal. Notable invertebrates present include dragon- and damselflies, notonectid and corixid bugs, and beetles. All of these, of course, can fly as adults. Notable absentees are many of the larger crustaceans, though some not known to have resistant or aerial stages may occur; the amphipod *Austrochiltonia* provides an example.

Fish are often absent, but many frogs, reptiles and birds are conspicuous vertebrate members of the fauna. Wetlands are particularly important as wildfowl habitats (chapter 7), and indeed this importance provides one of the strongest and most easily understood arguments for their conservation in Australia. Enormous numbers of waterbirds in addition to ducks may also be supported by individual wetlands.

Rain-pools and puddles

The most ubiquitous of all standing waters are rain-pools or puddles; small, shallow, and ephemeral bodies of fresh water formed by rain in suitable depressions (Fig. 2.7 (b), (c)). Any that persist for more than a few weeks develop a diverse and highly characteristic fauna. The most comprehensive study of such pools applies to a series near Melbourne and in eastern Victoria, and it is upon this study that the following brief account is based. Regional differences are unlikely to be great.

Salinities are low, but fluctuate markedly and may increase to over 3‰ shortly before pools dry. Turbidities, on the other hand, are generally high, and may restrict photosynthetic activity in the phytoplankton. Anoxic conditions do not occur, and water temperatures follow air temperatures within a few hours.

The composition of the phytoplankton is unknown, and the length of pool life appears too short for the development of large aquatic plants (but submerged terrestrial vegetation, chiefly grasses, may be present). The fauna is not easily divisible into plankton and benthos communities, but is diverse and characteristic. It comprises numerous species of microcrustacean (Cladocera, Copepoda, Ostracoda), as well as a few large crustaceans (Conchostraca, Anostraca, Notostraca). Various non-crustacean groups also occur: rotifers, oligochaetes, water mites, gastropods, dipteran larvae (culicids, chironomids, stratiomyids), notonectids, corixids, beetles, caddis- and mayflies. Although many of the species are also found in more permanant fresh waters, such as farm dams, a sizeable proportion of them appears to be restricted to rain-pools and typically exhibits a predictable pattern of succession. An interesting example of a characteristic rain-pool species is *Saycia cooki*, a cladoceran. This is one of the commonest species in Victorian rain-pools, yet until these were investigated was considered a very rare animal.

Other small bodies of fresh water

There are other small, more or less permanent bodies of fresh water not easily grouped with any type of standing water so far discussed. Included are permanently filled rock-holes (Fig. 2.7(d)), collections of water in cavities associated with terrestrial plants, natural impoundments of mound springs (Fig. 11.2(a)), cave pools, and pools in crayfish burrows. Many of the surface waters are widely distributed in arid regions, and were sources of water for aboriginal communities and European explorers. Space and, in some cases at least, a lack of adequate knowledge preclude proper consideration of them here. All are inhabited by a characteristic and often diverse fauna, with many members highly adapted to the vagaries of life in such water-bodies. Several references listed for this chapter provide further information.

3 Rivers and streams

Until quite recently, freshwater biologists paid far less attention to running waters than to lakes and other standing waters. There are many explanations for this, but it is nevertheless surprising, given the importance of rivers and streams to man. The Australian position reflected the international one, and until a decade ago studies of Australian rivers and streams were few indeed.

The present position is different. In recent years, many studies of running waters have been undertaken, particularly in North America and Europe. It has even been said, not without some justification, that river ecology has become one of the most exciting and active areas of modern scientific research. Several recent studies of Australian rivers and streams have also been made, but, for the most part, they have been largely (and necessarily) descriptive. Thus, the extent to which general concepts about the nature of running waters apply in Australia, these concepts being derived largely from northern hemisphere studies, has yet to be determined. Although it is likely that substantial applicability will be found, Australian running waters have a sufficient number of special features to promise investigators rich rewards at both regionally descriptive and globally conceptual levels. A consideration of these special features, as well as brief descriptions of the principal types of Australian river and stream habitats, forms the basis of this chapter. To provide perspective, the chapter begins by briefly considering some recent views about the broad ecological nature of running waters.

AN ECOLOGICAL PERSPECTIVE

Early ecological views of running waters were essentially that rivers and streams were drainage channels linking land and sea, without any ecological integrity of their own, and in which there was no significant processing or cycling of matter. In the last 20 years such views have become obsolete. Rivers and streams are now seen as integral parts (subsystems) of catchment ecosystems (drainage basins), with well-defined pathways and mechanisms for the processing and transport of substantial amounts of matter.

Correspondingly, river and stream studies have become less concerned with documenting the distribution and abundance of component plants and animals, and have stressed more the sources and fates of organic matter and inorganic nutrients. An expression of the results of this emphasis is provided by the so-called River Continuum Hypothesis, an attempt to summarize the major ecological attributes of a river system from its headwaters to the lowermost reaches.

According to this hypothesis, 'headwaters' are characterized by: (a) heterotrophy (energy consumed by community respiration exceeds energy fixed by gross photosynthesis) as a result of shading by bankside vegetation; (b) a dominance of 'shredders' and 'collectors' amongst the macroinvertebrates; and (c) a significant input of coarse particulate organic matter from terrestrial sources. 'Middle reaches' are characterized by: (a) autotrophy (community respiration is less than gross photosynthesis) because of reduced bankside vegetation and shallow and clear water; (b) a dominance of macroinvertebrate 'collectors' and 'grazers'; and (c) a decreased input of coarse particulate organic matter from terrestrial sources together with a significant input of fine particulate organic matter from upstream. The 'lowermost reaches' are characterized by: (a) heterotrophy because increased depth and turbidity depress photosynthesis; (b) a dominance of macroinvertebrate 'collectors'; and (c) a dependence on inputs of fine particulate organic matter from upstream.

It is as well perhaps to explain the terms shredders, collectors, scrapers and predators as applied to the macroinvertebrates. The 'shredders' are those species that feed by shredding coarse organic matter such as leaves and bark. They probably obtain as much energy from surface microbial material as from the plant material itself. Examples include certain mayfly nymphs and caddis larvae. 'Collectors' feed by filtering fine organic particles from the water or bottom materials. Mussels, blackflies and certain midge larvae provide examples. 'Scrapers' feed by removing algae and other organic material from submerged surfaces. Certain caddis larvae and snails are examples. Finally, 'predators' are the carniverous macroinvertebrates. Notable predators are dragonfly nymphs.

Associated with the River Continuum Hypothesis is the concept of nutrient spiralling, viz the idea that each part of a river system is dependent upon energy leaked from 'inefficient' upstream processes of energy collection, use and storage. Both the River Continuum Hypothesis and nutrient spiralling concept can be regarded as the modern replacement of former schemes of longitudinal zonation which stressed the importance of physicochemical factors as determinants of community structure without reference to energy sources. The applicability of these older schemes to Australian streams and rivers has often been questioned. How applicable is the River Continuum Hypothesis?

SPECIAL FEATURES

Two characteristics of Australia's natural environment have an overriding impact upon the nature of its rivers and streams: the overall climatic aridity, and the flatness of the continent. The effect of aridity is most obviously seen in the small total annual run-off, 350 km³, a value far exceeded by several individual rivers on other continents. Run-off, moreover, represents only 13% of total annual precipitation, the lowest fraction for any continent.

Another effect of the climate is seen in the marked seasonal and year-to-year variation in river and stream flows. Figure 3.1 illustrates this by reference to river heights of the River Murray at Euston. Even greater seasonal variation in flow rates is displayed by running waters in the monsoonal regions of the north and northeast, and greater year-to-year variation by rivers with more arid catchments than the River Murray. These features are illustrated by Figs 3.2 and 3.3. In monsoonal regions, rivers carry huge volumes of water in summer, yet may almost cease to flow in winter. In arid catchments, rivers may carry significantly higher volumes of water than the mean for years at a stretch, then significantly less for equally long periods.

Australia's topographic flatness means that many inland rivers and streams have low gradients: their rate of vertical fall is extremely slight. This is especially true of the Murray–Darling river system. Here, zones of erosion

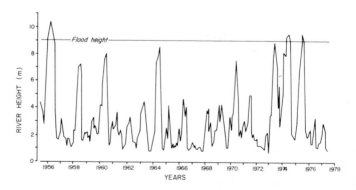

Fig. 3.1   River Murray heights (as a measure of river flow volumes) at Euston Weir, NSW, during period 1956–77. Redrawn after Pollard, Llewellyn and Tilzey (1980).

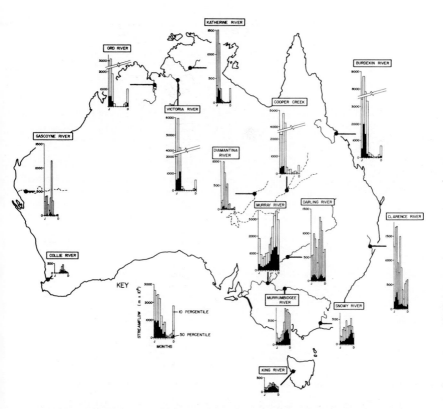

Fig. 3.2 Variability of monthly streamflow for selected rivers throughout Australia. The graphs show monthly streamflow volumes likely to be equalled or exceeded 1 year in 10 or 1 year in 2. Note different scales of vertical axes. Redrawn and modified from Department of National Resources, Australian Water Resources Council (1978).

(upper reaches) are short, and zones of deposition (lower reaches) long. The high turbidities of many inland rivers perhaps reflect this, though also implicated may be severe catchment erosion brought about by man's agricultural activities.

Two features are notable with regard to the chemical nature of rivers and streams. Firstly, the salinity of many Australian rivers (if not streams) is higher than the mean value for world rivers (120 mg/l) or the mean value for rivers on other continents (69–182). This is indicated in Fig. 3.4 where the mean salinity for all major river basins in Australia is plotted. A few Australian rivers have salinities which sometimes exceed 5000 mg/l! It has been claimed that

B

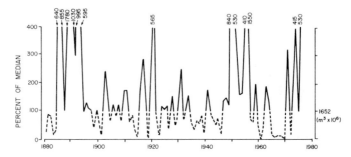

Fig. 3.3 Year-to-year variation in annual streamflow in the Darling River at Menindee, NSW. The median annual streamflow for the river at Menindee is 570 000 million m³. Redrawn from Department of National Resources, Australian Water Resources Council (1978).

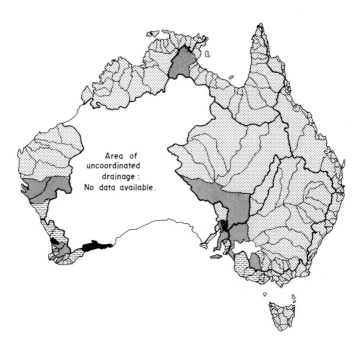

Fig. 3.4 Salinities of major rivers in principal basins. Map compiled by Dr K. F. Walker from data given by Department of National Resources, Australian Water Resources Council (1976). ░ fresh (salinity <500 mg/l); ■ marginal (salinity 500–1000 mg/l); ≣ mildly saline (salinity 1000–3000 mg/l); ■ saline (salinity >3000 mg/l).

Australian rivers are the most dilute of any continent (with an average value of 59 mg/l), but this claim is based on fewer and less reliable data than are now available. A more modern calculation could well show the reverse.

Secondly, the chemical composition of many Australian running waters differs from that regarded as typical except in arid areas. Many Australian rivers and streams provide exceptions to the so-called 'bicarbonate' composition in which calcium and magnesium bicarbonates are the most important salts. Frequently, sodium chloride is important, as in many lakes (chapter 2). Of course, there are also many rivers and streams which do conform to the 'bicarbonate' composition.

Amongst special biological features is the composition of the fauna. The special character of its various components is discussed at length in later chapters; it need only be recorded here that the composition of the fluviatile fauna as a whole is unmistakably Australian. It combines, amongst other features, high levels of endemicity, an absence of several groups common elsewhere, the development of great diversity in certain groups, and the presence of some groups unknown or uncommon elsewhere.

Another special biological feature relates to the nature of the terrestrial vegetation bordering temperate upland streams. As indicated above, litter (leaves, branches, bark) from this vegetation is an important source of energy for the fauna of headwater streams. In northern temperate regions, most of the litter arrives in a short and precisely timed period because the vegetation is deciduous in autumn. In Australia, most of the litter arrives over a long summer period. Australian litter also appears to be different in composition, and contain more bark and branches than northern temperate litter (mostly leaves).

The biological repercussions of these differences have yet to be explored fully. It has been suggested that one effect has been to induce imprecisely timed or flexible lifecycles. At least some groups of Australian stream invertebrates (stoneflies, mayflies) have such lifecycles. They have also been attributed to Australia's uncertain climate. Full exploration of this subject offers exciting prospects for Australian stream biologists. They should remember, however, to take account of the substantial differences between cool-temperate and warm-temperate streams in the northern hemisphere, and remember that most Australian upland streams are warm-temperate ones.

PRINCIPAL TYPES OF RUNNING WATER HABITATS

On the basis of climate, topography, and the distribution of the major endorheic, exorheic and arheic drainage basins (Fig. 3.5), five regions can be

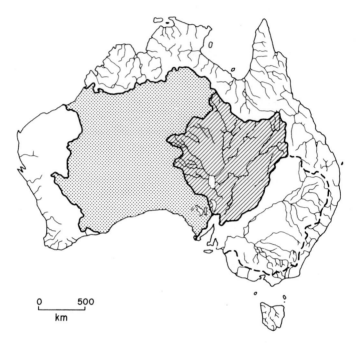

Fig. 3.5 Major drainage basin types in Australia. Note: in exorheic basins, most rain discharges to the sea via open lakes and river systems; in endorheic basins, drainage is internal, often to salt lakes; in arheic basins, significant surface-drainage patterns are absent. ░ Arheic drainage basins; ▨ endorheic drainage basins; ▢ exorheic drainage basins; – – – eastern boundary of Murray-Darling system.

distinguished in Australia within which running water habitats have some degree of similarity. The principal habitat types involved are: (a) the Murray–Darling river system; (b) upland rivers and streams; (c) northern flood-drought rivers; (d) rivers and streams east and south of the Great Dividing Range (and in parts of Western Australia); and (e) temporary streams. This simple categorization is, of course, an overgeneralization and should not be regarded as anything but a convenient framework upon which to build brief descriptive accounts of Australian rivers and streams. It is certainly quite inadequate as an expression of the total diversity of running water habitats in Australia. As for the individual accounts, the imbalances between them reflect the real differences which exist in our knowledge of the various sorts of Australian river and stream. Some accounts (e.g. of northern flood-drought rivers) rely largely on information from a single river; for others (e.g.

temporary streams) there is no comprehensive information on even a single locality.

Murray–Darling river system

A large area of inland southeastern Australia is drained by rivers which together constitute the Murray–Darling river system (Fig. 3.5). Perhaps of all inland waters in Australia, these rivers capture best the essential spirit of the 'outback' (Fig. 3.6). They drain over 1 million km², or one-seventh of Australia's landmass, but, apart from its eastern and southern flanks, the area is semi-arid and remarkably flat. In the main, therefore, most water originates in higher and wetter areas, and flows slowly across an immense dry plain with much meandering. Evaporative losses are high, and since input volumes are relatively low, the total flow in the system is remarkably small, only 22 km³. Only about half of this actually reaches the sea, and only about one-tenth of the total flow in the lower Murray is derived from the Darling. Thus, although one of the longest river systems in the world, the annual discharge is insignificant by world standards and its contribution to total run-off from the continent is less than 10%.

Whilst all the main rivers are permanent, marked seasonal variation in

Fig. 3.6 Murrumbidgee River.

streamflow characterizes the rivers in the southern part of the basin (Murray, Murrumbidgee, Lachlan), with maximum values in late winter/spring and minima in late summer/autumn (Figs 3.1, 3.2). Streamflows in the northern part show less well-developed seasonal patterns, though Darling headwaters in Queensland are fed by summer monsoons. Rivers throughout the basin show marked year-to-year variation (cf. Fig. 3.3).

Annual temperature ranges vary according to position; southern rivers (e.g. the Murray) have ranges from about 8–25°C, more northern ones from 10–30°C. Turbidities are almost invariably high. Salinities are mostly less than 500 mg/l, but in some southern rivers salinities can reach 3000 mg/l. The 'bicarbonate' type of chemical composition prevails in the fresh waters of the system.

Rather little is known about the biology of the system. We do know, however, that the fish fauna is remarkably depauperate by world standards; only 26 species of native fish occur. A few are endemic (e.g. the trout cod), but as a whole the fish fauna does not form an assemblage characteristic of the system. Nevertheless, many fish possess reproductive and behavioural responses of clear adaptive value. Thus, several have breeding seasons which coincide with flooding so that abundant food becomes available as waters inundate productive floodplain areas. Others produce large numbers of floating eggs which easily disperse over flooded areas.

The benthic fauna is likewise depauperate; in River Murray sediments it seems to consist largely of a few species tolerant to the unstable nature of the bottom and to fluctuating streamflows. Notable members are *Alathyria jacksoni*, a freshwater mussel, and *Euastacus armatus*, a crayfish. The greatest number of species is associated with submerged logs, branches and other tree litter rather than with sediments. Since the Murray, in particular, receives a huge litter input from river red gums, this habitat is a substantial one. It often contains dense populations of insects, crustaceans, molluscs and other animals clearly recognizable as shredders, collectors, scrapers and predators.

Because of the low rates of streamflow, the larger rivers of the system have a well-developed zooplankton community. At least some elements of this are endemic. In the largely unimpounded Darling, the zooplankton is basically a fluviatile one dominated by rotifers, but in the impounded Murray it is essentially a lacustrine community dominated by copepods and cladocerans (Fig. 3.7).

Similar differences are shown by the phytoplankton. In the Darling it is sparse, but is well-developed in the Murray, where cyanobacteria (*Anabaena, Anacystis*) dominate in summer and diatoms (*Melosira, Cyclotella*) in winter. Large aquatic plants are sparse in all rivers of the system, no doubt a function of fluctuating water levels and unstable sediments. However, associated with

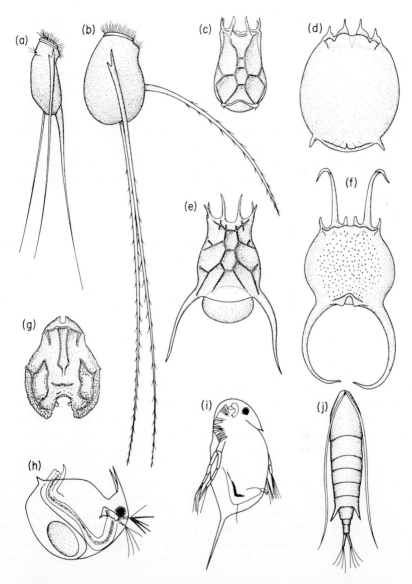

Fig. 3.7 Typical zooplanktonic forms in the River Murray. (a)–(g) rotifers; (h), (i) cladocerans; (j) copepod. (a) *Filinia pejleri*; (b) *Filinia australiensis*; (c) *Keratella shieli*; (d) *Brachionus calcifloris*; (e) *Keratella australis*; (f) *Brachionus falcatus*; (g) *Brachionous keikoa*; (h) *Bosmina meridionalis*; (i) *Daphnia carinata*; (j) *Boeckella triarticulata*. From Williams (1982) but redrawn after Shiel (1982 and original).

the rivers is a characteristic riverbank plant community, well-adapted to the hydrological regime. The most obvious trees of this community are the river red gum (*Eucalyptus camaldulensis*) and black box (*E. largiflorens*).

An extensive floodplain is also associated with the Murray and many of its Victorian tributaries. This is an integral part of these river systems; when not flooded, more or less cut-off parts of the river on it (billabongs) act as breeding and refuge areas for waterfowl, fish and other animals as well as sources of plankton to the river itself; when the floodplain is inundated, a rich source of terrestrial nutrients becomes available to the aquatic community. Billabongs are biologically much more diverse than their associated rivers, with complex macrophyte, planktonic, benthic and other communities present. They are considered in more detail in chapter 2.

Many rivers of the Murray–Darling system, particularly the Murray itself, have been greatly affected by man's activities, a subject considered at length in chapter 12. A major impact has been impoundment, with the result that floodplain rivers are now isolated from their floodplains. Ignorance of the effects of this isolation underlie many environmental problems associated with the rivers and their use.

Upland rivers and streams

The area within which fast-flowing rivers and streams occur is smaller than the area of the Murray–Darling system, though by no means inconsiderable in absolute terms. It comprises the Great Dividing Range stretching from north-eastern Queensland to western Victoria and a small area in southwestern Western Australia on the mainland, and a large part of Tasmania. There are of course pronounced climatic differences within such an extended area, but overall, the climate is wet and cool in the southeast and southwest, and wet and warm in the northeast.

Stream gradients are steep, so that the rivers and streams are eroding environments, not, as those of the Murray–Darling system, depositing ones. They are also permanent environments, and whilst seasonal fluctuations take place in streamflows, these are generally not pronounced. Flow rates are highest in summer in the northeast, and highest in winter/spring elsewhere. Turbidities and salinities are low. Seasonal temperature ranges vary according to position; in Victorian streams winter temperatures are often less than 5°C and summer temperatures about 20°C.

The speed of currents pre-empts the development of a plankton, but the worldwide phenomenon of drift, i.e. a diurnal pattern of suspended animals drawn from the benthic community, has been well-documented for a number of Australian upland streams. The general pattern of drift — in Australia as elsewhere — is for the numbers of drifting animals to increase during the hours of darkness with peaks just after sunset and before sunrise. There are, however, many variations from this pattern according to season, substratum type, species and so on. The significance of drift, in Australia as elsewhere, remains obscure, but its occurrence certainly fits the River Continuum Hypothesis.

As for the benthos itself, the number of species appears to correlate with the size of the stream or river and its catchment; small streams have fewer species than large ones. Nevertheless, species number as a whole (for southeastern localities at any rate) appears to be comparable to that of similarly sized upland streams and rivers of the northern hemisphere. Perhaps there are some differences in the relative abundance of certain groups (for example, lower than expected numbers of mayfly and stonefly species in southeastern streams), but it seems that increased numbers of species in other groups (caddisflies, beetles, chironomids) compensate. Some common and distinctive members of the benthic community of Australian upland streams and rivers are illustrated in Fig. 3.8. But readers are cautioned: there are many regional differences in faunas and a mere fraction of the total diversity is illustrated.

Northern flood-drought rivers

The northern tropical margins of the continent from Cape York Peninsula (Qld) to the Kimberleys (WA) are drained by numerous rivers which generally arise in upland areas but for most of their length flow across a wide coastal plain. Much of this plain is sparsely populated and seasonally inaccessible, and so it is not surprising that little is known about its rivers. However, one river, the Magela Creek (Fig. 3.9), a tributary of the East Alligator River some 250 km east of Darwin, has been extensively investigated in the past few years because of possible environmental damage by uranium mining. It is upon investigations of the Magela Creek that this account of flood-drought rivers in northern Australia relies.

Pronounced seasonal climatic differences dominate the character of the region (Fig. 3.2). There is a summer wet season from November to March, a winter dry season from May to September. In the wet season, vast amounts of water are discharged, often inundating huge tracts of the coastal plain in the process; the total annual amount discharged is 132 km^3, almost 40% of the total discharged by all Australian rivers. In the dry season, many of the rivers have negligible or no streamflow, and their floodplains are studded by a series of permanent standing water-bodies. These, though referred to as billabongs,

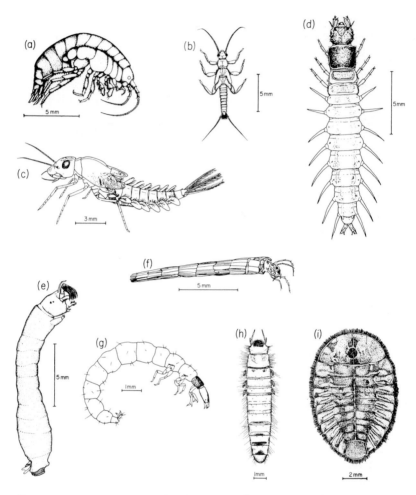

Fig. 3.8 Some common macroinvertebrates of upland streams and rivers. (a) Crustacea, (b)–(d) insect nymphs or larvae. (a) *Paramoera* (Amphipoda); (b) *Riekoperla* (Plecoptera); (c) *Mirawara* (Ephemeroptera); (d) *Archicauliodes* (Megaloptera); (e) *Simulium* (Simuliidae, Diptera); (f) leptocerid caddis in case (Trichoptera); (g) rhyacophilid caddis (Trichoptera); (h) helodid (Coleoptera); (i) psephenid (Coleoptera). Redrawn after various authors.

are unlike billabongs of the Murray–Darling system in many important respects, and are therefore better discussed here, not in chapter 2. Three types of billabong have been distinguished on the coastal plain: 'backflow billabongs' which receive backflow water from the main river in addition to

Fig. 3.9 Magela Creek, Northern Territory, during the dry season.

locally drained water; 'channel billabongs' located in the main drainage lines; and 'floodplain billabongs' filled only by spreading floodwaters.

Almost all features, biological and otherwise, are influenced by the seasonal hydrological picture (hence the term flood-drought rivers). Salinities fluctuate seasonally, with maxima occurring in the billabongs towards the end of the dry season. Overall, however, they are low. To some degree, the chemical composition of the waters also fluctuates seasonally, although generally the chloride and bicarbonate salts of sodium dominate throughout the year. Thermal stratification develops in the billabongs during the dry season, but is then more a diurnal phenomenon than a persistent seasonal one, with no stratification at night and an obvious one during the day (Fig. 3.10). High surface temperatures prevail throughout the year irrespective of any vertical stratification; they vary from about 22°C in July to over 40°C in November. Oxygen patterns reflect those for temperature. High concentrations and no stratification are the rule in the wet season; in the dry season, stratification in billabongs is more a diurnal phenomenon than a persistent seasonal one (Fig. 3.10). Note that oxygen concentrations are sometimes low in the bottom waters of deep billabongs during the dry season.

The biological communities of flood-drought rivers and creeks are diverse and often rich. The vegetation of the floodplain is dominated by grass, sedge and paperbark (*Melaleuca*) species, but near billabongs and drainage lines it typically includes pandanus palms (*Pandanus*), freshwater mangrove trees (*Barringtonia*) and further species of paperbarks (Fig. 3.9). Aquatic

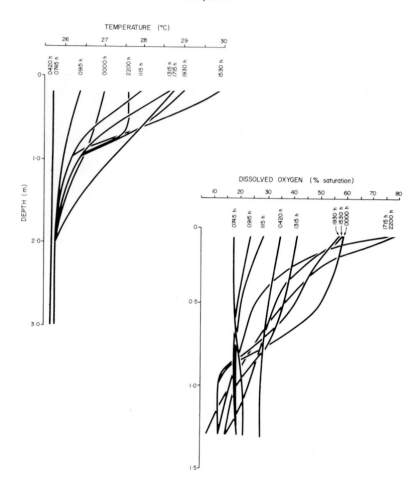

Fig. 3.10 Diurnal differences in the vertical pattern of temperature and oxygen in a billabong of the Magela Creek system, Northern Territory (Coonjimba, 25–26 April 1978). Note different vertical scales. Graphs redrawn from Walker and Tyler (1979).

macrophyte communities develop profusely in most billabongs during the wet season but gradually die with falling levels. Considerable damage to them is caused by the trampling and wallowing of buffaloes (see chapters 8 and 11), with consequential damage to other communities which use aquatic macrophytes mainly as either shelter (macroinvertebrates) or food (e.g. waterbirds, see Fig. 7.3). A wide variety of planktonic algae, likewise, has been recorded from Magela Creek billabongs. Increased concentrations of these develop during the dry season and often give rise to algal blooms. The algal flora strongly resembles that of the Indonesian archipelago.

Macroinvertebrate animals appear to be sparse in the deeper parts of billabongs, the only forms being freshwater mussels (*Velesunio angasi*), a palaemonid shrimp (*Macrobrachium rosenbergii*), oligochaete worms and various dipteran larvae (Chironomidae, Chaoborinae, Ceratopogonidae). Littoral macroinvertebrates are much more diverse, with both species diversity and abundance correlated with season: greatest diversity and abundance occur in the late wet to early dry season (April–July), and least at the end of the dry season. The commonest forms recorded from Magela Creek billabongs include triclads, gastropods, oligochaetes, hydracarines, and numerous crustacean and insect species. The lifecycles of at least some of these are very short, no doubt a reflection of the consistently high water temperatures; the lifecycles of some mayfly and caddisfly species seem to last no more than a month, with continuous breeding. Organic detritus is probably the main source of food for most of the non-predators. During the dry season the diversity of the zooplankton community is high in channel billabongs with up to 70 species of rotifer and microcrustaceans. Floodplain billabongs have lower diversity. The plankton fauna has many similarities to tropical plankton elsewhere and in particular to that of Indonesia.

The patterns of aquatic vertebrate distribution and abundance again strongly correlate with the hydrological regime. There is a diverse fish fauna (many more species than in the Murray–Darling system, for example), with breeding cycles which take advantage of periods of maximum food. During the wet season, wide dispersion of fish over the floodplain occurs (though fish 'kills' at the start of the wet season are frequently observed). Some species, notably the barramundi (*Lates calcarifer*, see chapter 5), migrate downstream to breed in estuaries.

Although frogs are common, the frog fauna is much less diverse; probably no more than about 20 species are found in the lowland reaches of flood-drought rivers. Apparently this is because breeding is severely limited by fish predation on eggs and tadpoles. Only *Litoria dahlii* is able to breed in billabongs, and other species breed in small ephemeral water-bodies mainly during the early wet season.

Little need be said about other vertebrates. A number of truly aquatic snakes occur (Fig. 6.6), both species of crocodile are found, and there is a rich diversity of waterbirds, some with pronounced migratory habits. Of the several species of freshwater tortoise present, the pitted-shell turtle (*Carettochelys insculpta*) is notable (see chapter 6).

Rivers and streams east and south of the Dividing Range (and in parts of Western Australia)

Numerous well-defined rivers and streams drain the regions east and south of the Great Dividing Range from northeastern Queensland to South Australia, and parts of southwestern Western Australia. Tropical to temperate in location, these running waters comprise the least homogeneous of the principal types distinguished in this chapter, and any attempt to draw a generalized picture would be contrived. There are, nevertheless, some common features.

All are more or less permanent, though seasonal hydrological differences may lead to reduced or negligible flows in some during the dry season (winter in the northeast, summer elsewhere). Most are large; rivers, not streams are the typical environments. Few are very long, so that easy seaward access is an obvious feature. And salinities are generally greater than those of other running waters, though still usually less than 500 mg/l.

Macroinvertebrate diversity is often high, and from some well investigated rivers (e.g. the Latrobe River system) over 300 species have been recorded. Diversity is least in lowermost reaches. Of the many animals recorded, special mention may be made of atyid shrimps, for they have planktonic larvae, an apparent anomaly for riverine animals. It seems that the shrimps survive by lifecycle adjustments. At least in the southeast, breeding coincides with periods of low flow and the occurrence of river pools (summer). In the case of more or less permanently fast-flowing environments inhabited by shrimps (*Atyoida striolata* in eastern rivers and streams), there is circumstantial evidence of breeding migrations to the estuary.

Fish diversity is relatively high, though differs regionally. It is highest in

Fig. 3.11 Temporary streams. (a) A pool in the otherwise dry bed of an episodic stream near Alice Springs in the Northern Territory; (b) an intermittent stream in the Flinders Ranges, South Australia.

the northeast (more than 50 species), and lowest in the southwest of Western Australia (less than 20 species). Many of the fish migrate between the freshwater and marine environments; eels, the Australian grayling and bass, and some native trout (galaxiids) provide examples.

Of all the principal types of running water in Australia, coastal rivers and streams are the most damaged by man. This general subject is addressed in chapter 12, and here only the high salinity now found in some coastal rivers is noted. This phenomenon is most obvious in southwestern Western Australia where it results from the clearing of catchments of native vegetation. In some rivers, the damage has been so great that upper reaches are more saline than lower ones, and river pools develop a salinity stratification in summer.

Temporary streams

Throughout the centre of Australia there are virtually no permanently flowing rivers or streams; flowing waters are temporary phenomena, appearing only after substantial rain. According to rainfall pattern, two types of temporary rivers and streams can be distinguished: 'episodic' (or 'ephemeral') in areas where rainfall has no well-defined seasonal pattern (more arid areas), and 'intermittent' ones where rainfall shows seasonal periodicity (less arid areas) (Fig. 3.11). Despite the huge area of Australia drained by temporary rivers and streams (over half of the continent), studies of these environments are extremely few. There is none of episodic streams. A few recent studies of intermittent streams in South Australia provide at least some insight into the composition of the macroinvertebrate fauna of this sort of temporary aquatic locality.

One locality studied, Brownhill Creek near Adelaide, is intermittent in its upper reaches; these flow only from winter to midsummer. When not flowing, the stream dries to a series of isolated pools. During streamflow periods, the fauna is dominated by immatures of chironomids, stoneflies, and a caddisfly species (*Leptorussa darlingtoni*). Beetles, especially dytiscids and noterids, are also present, though not abundant. Mayfly and dragonfly nymphs, molluscs, and crustaceans are notably rare or absent. When flow ceases, caddisfly larvae and beetles dominate the pool fauna.

A fauna of quite different composition occurs in intermittent streams in the Flinders Ranges, south of Lake Eyre. In one stream sampled when flowing (Brachina River), mayflies, dragonflies, and several dipteran families in addition to the Chironomidae were common in the surface waters, but stoneflies and cased caddisfly larvae were not. Still further groups were found in stream interstitial water, and in a hole dug 1 metre from the edge of the stream,

representatives of 14 major animal groups were recovered, including amphipods, isopods (janirids), stoneflies, molluscs and oligochaetes.

Several suggestions can be advanced to explain the differences between
the two localities. Undoubtedly involved is the fact that in one stream, it is the
upper reaches that are intermittent, and the lower reaches in the other. The
better refuge provided by the stony bottom of Brachina River is also no doubt
involved.

Even less is known about other elements of the fauna. The frog fauna is
apparently sparse, and the occurrence of frogs near temporary rivers and
streams seems to be governed largely by the presence of permanent pools and
the availability of nearby terrestrial refuges for adults during the dry season,
especially refuges provided by holes in the stumps of broken limbs of the
coolibah tree (*Eucalyptus microtheca*). The fish fauna, likewise, is not diverse;
although 22 native species are recorded from river systems of central
Australia, probably most records are of fish in riverine pools, not flowing
waters. One fish known to occur in flowing temporary waters is the mosquito
fish, an introduced species (see chapter 5).

4 Invertebrates

The overwhelming majority of animals are invertebrates and most of these are arthropods (insects, crustaceans and their allies). That statement applies generally, irrespective of region or habitat, but applies in particular to the fauna of Australian inland waters. The recent publication of *Australian Freshwater Life: The Invertebrates of Australian Inland Waters* (Williams 1980) provides a more detailed account than is possible here, and readers particularly interested in this part of the Australian aquatic biota and disappointed by the brevity of the present chapter are referred to that book.

Some general comments on the overall nature and origins of the aquatic invertebrate fauna are, on the other hand, quite appropriate, as are comments on groups of particular interest, on problems associated with invertebrates, and on the practical uses of some invertebrates. These topics provide the framework for this chapter. Beforehand, one introductory remark cannot escape attention, namely, that knowledge of Australian freshwater invertebrates is still meagre. Table 4.1 indicates just how meagre our taxonomic knowledge is. The table is arbitrary (especially in so far as it does not

Table 4.1 Status of taxonomic knowledge of Australian aquatic invertebrates. This table updates an earlier one (Williams 1976).

Invertebrate groups in Australian inland waters	Status of taxonomic knowledge*
Protozoa (Fig. 4.1(a))	+
Porifera (sponges)	+ + +
Cnidaria (coelenterates, hydras, jellyfish; Fig. 4.1(b))	+
Platyhelminthes: Temnocephalidea (Fig. 4.1(c))	+ +
Turbellaria (flatworms; Fig. 4.1(d))	+ +
Nemertea, Nematoda, Nematomorpha (gordian or horsehair worms), Gastrotricha, Tardigrada (Fig. 4.1(e))	+
Rotifera (wheel animalicules; Fig. 4.1(f))	+ +
Polyzoa (pipemoss, moss animalicules)	+ +
Mollusca: Bivalvia (mussels, pea-shells; Fig. 4.1(g))	+ +
Gastropoda (snails; Fig. 4.1(h))	+ +

Table 4.1 (Contd.).

Invertebrate groups in Australian inland waters	Status of taxonomic knowledge*
Annelida: Polychaeta	+ +
Oligochaeta (true worms; Fig. 4.1(i))	+ +
Hirudinea (leeches)	+ +(+)
Hydracarina (water-mites)	+ +
Crustacea: Anostraca (fairy-shrimps; Fig. 4.2(a))	+ + +
Notostraca (tadpole-shrimps; Fig. 4.2(b))	+ + +
Conchostraca (clam-shrimps; Fig. 4.2(c))	+
Cladoceran (water-fleas; Fig. 4.2(d))	+ +
Ostracoda (seed-shrimps; Fig. 4.2(e))	+ +
Copepoda: Calanoida (Fig. 4.2(f))	+ + +
Cyclopoida (Fig. 4.2(g))	+ +
Harpacticoida (Fig. 4.2(h))	+ +
Branchiura (fish-lice; Fig. 4.2(i))	+
Malacostraca: Syncarida (Fig. 4.3(a), (b))	+ +
Isopoda (Fig. 4.3(c), (d), (e))	+ +
Amphipoda (shrimps; Fig. 4.3(f))	+ +
Atyidae (shrimps, prawns; Fig. 4.3(g))	+ +
Palaemonidae (shrimps, prawns)	+ +
Parastacidae (crayfish, marron; Fig. 4.3(h))	+ +
Hymenosomatidae (crabs; Fig. 4.3(i))	+ + +
Sundathelphusidae (crabs)	+ + +
Insecta: Collembola (spring-tails)	+ +
Plecoptera (stoneflies; Fig. 4.4(a))	+ + +
Ephemeroptera (mayflies; Fig. 4.4(b))	+ +
Odonata (dragonflies, damselflies; Fig. 4.4(c), (d))	+ + +
Hemiptera (bugs, backswimmers, etc; Fig. 4.4(e))	+ +(+)
Megaloptera (alderflies; Fig. 4.4(f))	+ + +
Neuroptera (lacewing flies)	+ +
Mecoptera (Fig. 4.4(g))	+ +
Diptera: Blephariceridae	+ + +
Simuliidae (blackflies; Fig. 4.4(h))	+ +
Culicidae (mosquitoes, midges; Fig. 4.4(i))	+ + +
Chironomidae (gnats, midges; Fig. 4.4(j))	+ +
Other Diptera (Dixidae, Tipulidae, Ceratopogonidae, Ephydridae, etc.)	+ +
Hymenoptera (parasitic wasps)	+
Lepidoptera (moths)	+
Trichoptera (caddisflies; Fig. 4.4(k))	+ +(+)
Coleoptera (beetles; Fig. 4.4(l), (m))	+ +

* +, little taxonomic information available; + +, not well-worked taxonomically; + + +, relatively well-worked taxonomically, at least in part (indicated by parentheses). The information relates to *aquatic* stages only.

distinguish between regions and life-history stages), but its message is quite definite: there are a great many invertebrate groups for which little taxonomic information is available or which are not well-worked taxonomically.

Since Table 4.1 lists all aquatic invertebrate groups occurring in Australia, and, where applicable, gives common names, it may provide a useful reference for readers unfamiliar with the unavoidable plethora of scientific names in this chapter. Simple line drawings (Figs 4.1–4.4) illustrate most of the groups referred to in the table. They have been prepared to provide non-zoological readers with an indication of the diversity of form shown by freshwater invertebrates and a visual guide to the major groups. Readers familiar with invertebrate types are merely offered a little visual relief. Some other invertebrates are illustrated in Figs 3.7 (rotifers, cladocerans, copepods) and 3.8 (an amphipod and various insect nymphs or larvae).

GENERAL COMMENTS

In overall composition, the invertebrate fauna of Australian lakes and rivers is not grossly dissimilar to the aquatic fauna of other continents. All the major freshwater groups occur, and within analogous environments the basic elements of the invertebrate fauna are similar to those elsewhere; by way of example, insects dominate stream faunas (Fig. 3.8), copepods, cladocerans and rotifers comprise the zooplankton of standing waters, and oligochaetes and chironomid larvae are important in the benthic (bottom) fauna of lakes (Table 2.4). It is when the composition is closely examined that several quite distinctive features become apparent.

Perhaps the most obvious feature is the high degree of endemicity. Endemicity is of course most pronounced in groups not easily dispersed, but it is also a feature of some groups which *a priori* would seem easily dispersed.

At the species level it is high in all groups of insects and crustaceans; these have many endemic species or their species are mostly endemic. Endemicity of this sort is also high in the temnocephalids (small animals ectocommensal on various freshwater crustaceans), turbellarians and molluscs. Groups with fewer endemic species, but nevertheless with some are the Porifera, Rotifera, Annelida, and Nematomorpha. The Protozoa, Cnidaria (one family excepted), Nemertea, Gastrotricha, Polyzoa and Tardigrada—in short, many of the 'lower' invertebrate groups—appear at this stage in our knowledge to have no or few endemic species. Note, however, that ideas concerning the level of species endemicity in Australian aquatic invertebrates are currently undergoing revision. Until a few years ago, most Australian rotifer and cyclopoid copepod species in particular were regarded as cosmopolitan or at least ubiquitous forms; recent work now indicates much higher levels of species endemicity than were thought to exist. It may well be that rigorous examination of certain other groups will reveal a similar picture.

Endemicity is less pronounced at the level of the genus. Nevertheless, many genera of the Syncarida, Isopoda, Plecoptera and Odonata are endemic, and there are several endemic genera in the Temnocephalidea, Turbellaria, Mollusca, Annelida, Hydracarina and remaining crustacean and insect groups. Predictably, endemicity is even less pronounced at the family level; however, the Syncarida has three endemic families (Anaspididae, Koonungidae, Psammaspididae), the Odonata two (Lestoideidae, Hemiphlebiidae), and the Trichoptera one (Plectrotarsidae).

Another distinctive feature of the Australian aquatic invertebrate fauna is that it lacks several groups elsewhere widespread, and typical or common in aquatic environments. The freshwater mussel families Mutelidae, Unionidae and Margaritiferidae are absent, as also are the fish leeches (Piscicolidae), certain cladoceran families (Leptodoridae, Holopedidae, Polyphemidae), freshwater isopods of the Asellidae (particularly of *Asellus*), mayflies of the family Ephemeridae, and representatives of the important caddis families Phryganeidae, Lepidostomatidae and Molannidae (and even the Limnephilidae has only a few Australian representatives).

A third distinctive feature of a general sort is the adaptive radiation within Australia of several families which elsewhere are not notably diverse. Crustacean families exhibiting this feature include the Centropagidae (calanoid copepods), Amphisopidae and Phreatoicidae (isopods); insect families are the Leptophlebiidae (mayflies), and Rhyacophilidae, Hydropsychidae and Leptoceridae (caddisflies). All these families have large numbers of species in Australia, many occupying particular habitats elsewhere more characteristically occupied by representatives from other families. A good example is provided by the leptophlebiid genus *Jappa* (Fig. 4.4(b)). This burrows in mud and its gills, forelegs and headtusks invite ready comparison with members of the Ephemeridae which also burrow. The Ephemeridae, as noted, does not occur in Australia.

In a sense, the diverse origins of the Australian freshwater invertebrate fauna may represent yet a further distinctive feature. There is a strong contingent of old freshwater forms of Gondwanaland origin, i.e. forms derived from inhabitants of fresh waters of the former and now split, southern hemisphere landmass comprising Antarctica, parts of Africa, Madagascar, South America, India and the whole of Australasia. This contingent is particularly strong in southeastern Australia (including Tasmania), and many members have a distribution in the southern hemisphere reflecting their origin. Examples are provided by the Hyriidae (the only Australian freshwater mussel family), some cladoceran species (e.g. *Chydorus eurynotus*), many ostracod species, centropagid and at least some harpacticoid copepods, some bathynellaceans (Syncarida), phreatoicid isopods, and parastacid crayfish. Outside Australia all of these are found either in New Zealand, New Guinea, the tip of South Africa, Madagascar, South America, and India, or in a combination of these places. Additionally, the cool-adapted stream insect fauna of southeastern Australia (species of Ephemeroptera, Plecoptera, Mecoptera, Trichoptera, Diptera), in particular, includes a number of amphinotic (circum-Antarctic) groups with close relatives in New Zealand and South America.

A second strong contingent comprises forms derived from the fresh waters of Asia by southward migration through the Indonesian archipelago. Examples of this oriental group are provided by many dragonfly genera, some atyid prawn genera (*Caridina, Paratya*) (Fig. 4.3(g)), and crabs of the Sundathelphusidae.

A third important contingent comprises forms which have originated in fresh waters of the northern hemisphere and conterminous landmasses and which have been carried to Australia by winds, on the feet of migratory waterbirds, or by other means. Examples, of course, are those groups with low levels of endemicity and of ubiquitous or cosmopolitan distribution (e.g. protozoans, many rotifers).

A fourth contingent is of marine origin. This is less important than the previous contingents but is perhaps relatively more important in Australia than in other continents. At least two waves of emigration from the marine environment can be distinguished: an old one whose representatives now occur in quite fresh waters and often well away from the coast, and a more recent one whose representatives are usually found not far from the coast and in fresh to slightly saline waters. Probable examples of the first emigrant wave are the janirid isopods, hydrobiid snails, and the prawn *Palaemonetes*. Of more recent emigrants, probable examples are the endemic coelenterate family Australomedusidae, isopods belonging to the families Sphaeromatidae, Cirolanidae and Anthuridae, corophiid amphipods, and hymenosomatid crabs (*Amarinus*) (Fig. 4.3(i)).

Finally, reference is made to a fifth and still less important contingent. This comprises forms that have evolved from local terrestrial ancestors. At least one good example is known, *Haloniscus* (Fig. 4.3(d)), an oniscoid isopod inhabiting Australian salt lakes. Its ancestors were almost certainly terrestrial.

SPECIAL GROUPS

Nine forms in three categories have been selected for discussion:
1 non-arthropods: Temnocephalidea, *Stratiodrilus* (a polychaete), freshwater mussels;
2 crustaceans: *Anaspides, Haloniscus,* phreatoicids;
3 insects: stoneflies, Nannochoristidae (Mecoptera), *Philanisus* (Trichoptera).

   In selecting these, criteria borne in mind were the possession of obvious adaptations to the nature of the Australian inland aquatic environment, the display of marked endemicity, primitiveness or diversity, or the occurrence of a pronounced amphinotal distribution (criteria sometimes, of course, in combination). However, *Philanisus* was selected because of its bizarre habits, and the Temnocephalidea and *Stratiodrilus* additionally in the hope that, by drawing attention to them here, more work will be stimulated on these little-known but fascinating members of the Australian fauna. Many different forms could have been selected using the same criteria.

   The Temnocephalidea and *Stratiodrilus* have many features in common. Both are ectocommensals or ectoparasites of freshwater crustaceans, both are little-known biologically and have had chequered careers with regard to their zoological affinities, and both are mainly (Temnocephalidea) or entirely (*Stratiodrilus*) confined to the southern hemisphere. Both deserve to be better studied—the Temnocephalidea in particular, for it is neither rare nor, within Australia at least, of restricted distribution; there are 15 known Australian species, and the class is recorded from all states except the Northern Territory.

   The Temnocephalidea (Fig. 4.1(c)) were first thought to be annelids, but were later recognized as Platyhelminthes. Exactly what sort of platyhelminth has been a subject of debate. At present they are perhaps best regarded as a separate and distinct platyhelminth class. Small stumpy animals, 1–12 mm long, with from two to six anterior tentacles, they are found living ectocommensally on a variety of freshwater crustaceans in Australia (many genera of crayfish, *Phreatoicopsis*, atyid and palaemonid prawns, sundathelphusid crabs). They occur attached to the outer body surface of hosts or, more often perhaps, within the branchial chamber formed by the lateral lobes of the carapace. Food includes small crustaceans, certain insect larvae, rotifers and nematodes. At least one species also feeds on algae. The class is most abundant in the southern hemisphere (Australia, New Zealand, Papua New Guinea, Madagascar, South America), but is known to exist in India, parts of Europe, Central America, and islands of the Far East. Note that despite this wide distribution, few textbooks dealing with the freshwater biota mention this interesting but largely forgotten group.

   The diversity of crustacean hosts of *Stratiodrilus* (Fig. 4.5(a)) is more

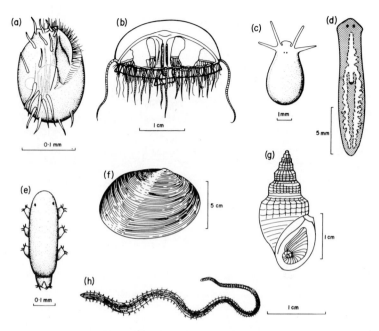

Fig. 4.1 Some invertebrates of Australian inland waters other than crustaceans and insects. (a) *Euplotes* (Protozoa); (b) *Craspedacusta sowerbyi* (Cnidaria); (c) *Temnocephala quadricornis* (Temnocephalidea); (d) *Dugesia* (Turbellaria); (e) *Macrobiotus* (Tardigrada); (f) *Velesunio* (Bivalvia); (g) *Plotiopsis* (Gastropoda); (h) *Tubifex* (Oligochaeta). Redrawn after various authors.

restricted; only parastacid crayfish are known hosts. On these, *Stratiodrilus* occurs in the gill chambers. The genus is a member of the primitive annelid family Histriobdellidae, one of only a few polychaete families that have been successful in penetrating the freshwater environment. About 1 mm long, the animal bears on its head a median tentacle, two pairs of lateral tentacles, and a pair of lateral palps. The body is 5-segmented, and the tail region imperfectly segmented but with two limblike structures possessing cirri, and, in males, claspers. Two Australian species have been described, and the genus is recorded from Queensland, New South Wales and Tasmania. Outside Australia, it has been recorded from South America and Madagascar. The only other genus in the family, *Histriobdella*, is associated with marine lobsters in the northern hemisphere.

The mussel family, Hyriidae (e.g. *Velesunio*, Fig. 4.1(f)), is discussed because of its amphinotic distribution (Australasia and South America), high level of endemicity and diversity (17 known Australian species, 15 endemic),

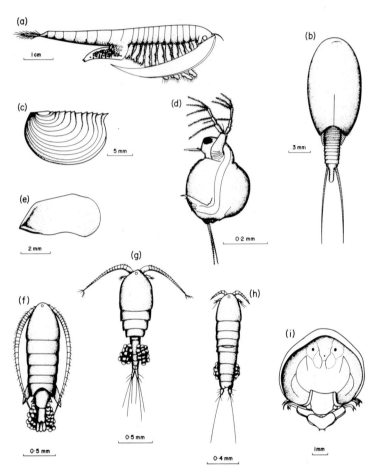

Fig. 4.2 Australian freshwater entomostracan Crustacea. (a) *Branchinella* (Anostraca); (b) *Lepidurus* (Notostraca); (c) *Limnadopsis* (Conchostraca); (d) *Moina* (Cladocera); (e) *Mytilocypris* (Ostracoda); (f) ovigerous female of *Boeckella* (Copepoda, Calanoida); (g) ovigerous female of *Microcyclops* (Copepoda, Cyclopoida); (h) ovigerous female of *Attheyella* (Copepoda, Harpacticoida); (i) *Dolops* (Branchiura). Redrawn after various authors.

and because several features of its biology are of interest, and some, obvious adaptations to the fluctuating nature of the Australian inland aquatic environment. The ability of adults of certain species (*Velesunio wilsonii*) to withstand prolonged desiccation, sometimes for years, is a clear example of this sort of adaptation. Other interesting features of the biology of the group include the nature of the lifecycle, the *relative* tolerance of at least some species to high levels of salinity, and the pattern of ectoparasitic mite infestation. Most, if not all, freshwater mussel lifecycles involve the production of a small larva, a glochidium, which lives ectoparasitically on freshwater fish. Little is known of the details of Australian lifecycles, but recent work has indicated that a number of species of native fish may act as hosts, and at least in the laboratory even tadpoles may oblige! Since many native fish disperse widely within river systems, especially during floods, this leads to wide dispersal of glochidia. Although the distributions of freshwater mussels have been used to uphold the concept of fluvifaunular provinces in Australia, viz the division of the country into regions according to river systems characterized by particular assemblages of freshwater animals, recent reanalysis has shown that mussel distributions provide little or no support for the concept. With regard to salinity tolerance, note that some of the most widely spread southeastern species can tolerate salinities up to 20°/oo for periods of 5 days, and in the field are found at salinities up to 3.5°/oo. In species investigated so far, *Alathyria jacksoni* and *Velesunio ambiguus*, osmoregulation occurs at low salinities, but animals osmoconform at higher salinities. Finally, brief mention is made of mite ecto-parasites. Species of the hydracarine genus *Unionicola* are frequently found living in the mantle chamber of host mussels, sometimes several different species within a *single* host, probably each with its own territory (though this remains to be confirmed).

By any set of criteria, *Anaspides* (Fig. 4.3(b)) must rank as one of the most interesting of all freshwater crustaceans. Discovered first in 1893 by a visiting New Zealand scientist, G. M. Thomson, it was initially allied with the mysid shrimps. Subsequent investigation soon showed that its affinities were in fact with certain fossil Palaeozoic Syncarida. The modern classification of this group involves three orders, only two of which are still extant, the Bathynellacea and the Anaspidacea. The former comprises small, widespread forms specialized for subterranean life and with many aberrant features. The Anaspidacea, on the other hand, contains relatively unspecialized primitive forms confined now to Australia (all families) or New Zealand and South America (stygocarids only). In general terms, *Anaspides* can be regarded as the most primitive true malacostracan known, and is thus of great interest to zoologists because it illustrates the basic pattern (the so-called caridoid facies) from which all higher crustaceans probably evolved. Fossils similar to *Anaspides* are known from Triassic strata some 180 million years of age. Two primitive morphological features are the absence of a carapace and the similarity between abdominal and thoracic segments and appendages. Another primitive feature is that eggs are not brooded but simply shed into the water where they undergo direct development to adults without the intervention of

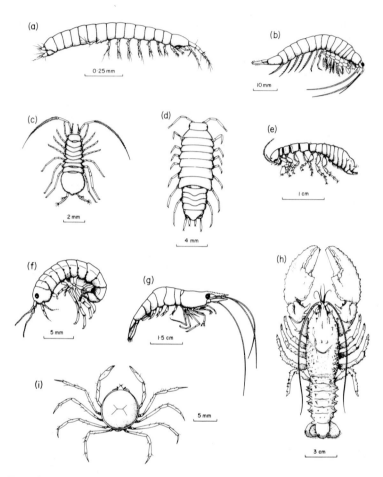

Fig. 4.3 Australian freshwater malacostracan Crustacea. (a) *Bathynella* (Syncarida, Bathynellacea); (b) *Anaspides tasmaniae* (Syncarida, Anaspidacea); (c) *Heterias* (Isopoda, Janiridae); (d) *Haloniscus searlei* (Isopoda, Oniscidae); (e) *Metaphreatoicus australis* (Isopoda, Phreatoicidea); (f) *Austrochiltonia* (Amphiopoda); (g) *Paratya australiensis* (Decapoda, Atyidae); (h) *Euastacus armatus* (Decapoda, Parastacidae); (i) *Amarinus lacustris* (Decapoda, Hymenosomatidae). Redrawn after various authors.

larval stages. There are two species, both known only from Tasmania. *A. tasmaniae*, still relatively common, is found in upland streams and creeks. Formerly, it seems, lakes were also inhabited, but introduced trout have ousted it from these. *A. spinulae* is known only from Lake St Clair. Individuals of *A. tasmaniae* live for 3–4 years and may reach lengths of 5.5 cm.

The diet comprises both animal and plant material. Several other anaspidaceans have recently been found in both Tasmania and continental Australia; none, however, arouses quite the interest of *A. tasmaniae*.

Haloniscus (Fig. 4.3(d)), a distant relative of the garden slater, is a genus of isopod of which the principal Australian species, *H. searlei* (one of two), lives completely submerged in salt lakes. The genus is endemic to Australia, and with the possible exception of *Desertoniscus birsteini* found in Turkmenia, USSR, *H. searlei* is the only known oniscoid isopod to live in inland salt water. It is widespread in suitable localities in southwestern, southern and southeastern Australia. There is strong evidence that the species is secondarily aquatic, and evolved from terrestrial ancestors. Within salt lakes it is known to descend to depths of 14 m, and to occur within a salinity range of 3.6–192°/°° (about one-tenth to five times the salinity of the sea). Not surprisingly, it is a powerful haemolymph osmoregulator and can regulate the osmotic pressure of its internal body fluid both well above and below low or high external salinities. In this ability it approaches closely the ability of the much better known salt lake crustacean, *Artemia* (an introduced animal in Australia). The recent claim that *H. searlei* is amongst the most remarkable of crustaceans living in hypersaline inland waters seems well-justified on a number of grounds.

The Oniscoidea, to which *Haloniscus* belongs, is of worldwide distribution, as indeed are most of the nine suborders of the Isopoda. The suborder Phreatoicidea (Fig. 4.3(e)), however, is found only in Australasia, India and the tip of South Africa. But it is within Australia that its greatest development occurs in terms of species diversity and abundance; there are 17 genera and many more species, all endemic. Known from all states except Queensland, in some areas phreatoicids are the commonest or only isopods present (the notable absence of the freshwater isopod family Asellidae in Australia has already been mentioned). There are several morphological features which distinguish them from other isopods, the most obvious being a lack of dorso-ventral flattening. Many different habitats are occupied. Included are springs, hot salty bore water, and damp soil, as well as typical freshwater lake and stream environments. Obviously the group as a whole is a successful element of the Australian freshwater fauna, and has been well able to cope with the rigours of the Australian aquatic environment over long periods.

Of all Australian aquatic insects, the Plecoptera (Figs 3.8(b), 4.4(a)) or stoneflies are the most typical in the sense that so many of the features which distinguish freshwater insects in Australia are highlighted in this order. The order is of undoubted antiquity (fossils are known from the Permian). All four Australian families display amphinotic ditributions, and, indeed, the close relationships between New Zealand, South American and Australian stoneflies

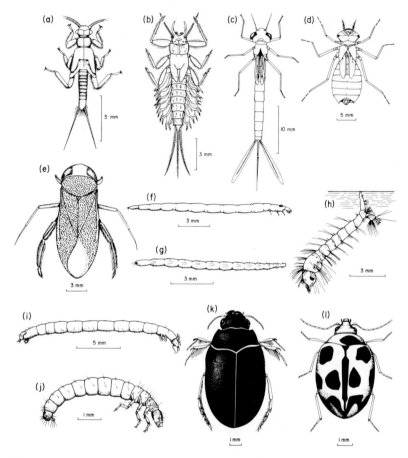

Fig. 4.4 Australian freshwater insects. (a)–(d), (f)–(j) are nymphal or larval forms; (e), (k) and (l) are adults. (a) *Dinotoperla fontana* (Plecoptera); (b) *Jappa* (Ephemeroptera); (c) *Caliagron billinghursti* (Odonata); (d) *Hemicordulia australiae* (Odonata); (e) *Agraptocorixa eurynome* (Hemiptera, Corixidae); (f) *Nannochorista* (Mecoptera); (g) *Culicoides* (Diptera, Ceratopogonidae); (h) culicid (Diptera, Culicidae); (i) *Chironomus* (Diptera, Chironomidae); (j) *Agapetus* (Trichoptera); (k) *Hygrobia* (Coleoptera); (l) *Macrohelodes* (Coleoptera). Redrawn after various authors.

are amazing (it is no coincidence that stoneflies were amongst the first groups whose distribution and internal affinities were used to support the concept of continental drift when this was still largely in the process of being accepted by the scientific community). Endemicity is complete at the species level (involving some 70 species), and almost so at the generic level (24). And many species appear to have more flexible lifecycles and longer flight periods than stoneflies on other continents, a phenomenon regarded as an adaptation that evolved in response to the uncertain nature of the Australian climate: Australian stoneflies, as a result, do not exhibit the sequential growth and flight periods characteristic of many closely related species in the northern hemisphere. These points are not, by any means, an exhaustive list of stonefly features of interest to biologists.

The order Mecoptera (scorpionflies) is of interest on quite different grounds. Outside Australasia (and a small area of South America) it is terrestrial. There is, however, one family, the Nannochoristidae, with aquatic larvae in Australia, New Zealand and South America. These (Fig. 4.4(f)) are long slender forms with biting mouthparts, strong legs and a terminal pair of anal hooks. For many years the aquatic habit was only suspected, but recent collections confirm the unequivocally aquatic nature of the larval habitat. Adults frequent swampy areas, lakeside margins and streams in upland regions of Tasmania and the southeastern part of the Australian mainland.

Finally for insects, reference is made to *Philanisus plebeius* (Fig. 4.5(b)), a caddisfly (Trichoptera). Strictly, a discussion of this species is misplaced since

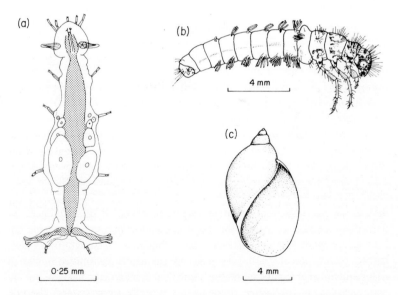

Fig. 4.5 Some interesting or economically important aquatic invertebrates. (a) *Stratiodrilus* (Annelida); (b) *Philanisus plebeius* (Trichoptera; (c) *Austropeplea tomentosa* (Gastropoda). Redrawn after L. Harrison, M. Quick and R. Plant.

the larvae occur in intertidal rock pools. However, because almost all Trichoptera inhabit fresh waters, the habitat of *P. plebeius* for the purposes of this account may be regarded as no more than a somewhat special sort of water-body associated with land — albeit strongly influenced by marine events! The highly distinctive nature and uniqueness of the species and its biology give additional justification for considering it here. *P. plebeius* belongs to the Chathamiidae, a small family of caddisflies whose larvae are marine and occur in the Australasian region. *P. plebeius*, itself, is found on the coasts of New South Wales and New Zealand. Although caddis larvae are known elsewhere from localities subject to marine influences (*Limnephilus affinis*), it appears that *P. plebeius* and the other few species of the Chathamiidae are the only truly marine caddises. But the interest of the species does not end there: the female has the unique habit of ovipositing in the coelom of a starfish, *Patirella exigua*, and, unlike other marine insects (e.g. Hemiptera), the larval cuticle is approximately as permeable as that of freshwater insects. Osmoregulation studies have shown, predictably, that salts in the haemolymph are strongly regulated. The larvae feed on calcareous algae, also used to build the larval case, and soft algae. Adults are restricted to coastal situations where they are mostly associated with rocky terrain.

PROBLEMS

It would be surprising if man were not in conflict at one time or other with some of the vast number and diversity of aquatic invertebrates. There are, of course, many problems associated with aquatic invertebrates, and we are lucky in Australia that some of the most significant (particularly those of tropical fresh waters) are largely absent.

In very broad terms the major problems in Australia relate to diseases of man and animals of economic importance, and involve organisms aquatic for the whole or part of their life. Also involved are several insect groups whose aerial phases though not medically important are nevertheless significant pests. Often the aquatic organisms implicated are not directly responsible for the disease but are important as either vectors (carriers) of the pathogen, intermediate hosts, or are part of the lifecycle of vectors or hosts. Discussion can most conveniently follow a systematic path, though not a straight one. Omitted from discussion are pathogens which are frequently transmitted to man in water and are able to survive free for some time in an aquatic medium but are best regarded as freshwater contaminants (usually from sewage). Several pathogenic bacteria provide examples: *Vibrio cholerae* (causing cholera), *Salmonella typhi* (typhoid), *Leptospira* spp. (leptospirosis). Briefly referred to are invertebrates causing problems to animals other than man and stock.

A protozoan genus of some concern is *Naegleria*, an amoeba-like organism found in damp areas. Most of its species are free-living and non-pathogenic, but some may produce pathogenic strains which may be transported in water. *Naegleria gruberi* and *N. aerobia* (*N. fowleri*) both produce pathogenic strains causing amoebic meningo-encephalitis, a usually fatal disease. Fortunately, not many cases occur in Australia, but a few occur each year, particularly during the warmer months (e.g. in South Australia).

The platyhelminth worms provide many important pathogens. Three are of particular medical significance, all in the genus *Schistosoma*: *S. haematobium, S. mansoni* and *S. japonicum*. Fortunately, they are insignificant in Australia, although several incidences of at least two of them are regularly recorded each year. They cause schistosomiasis or bilharzia, a severely debilitating disease common to most tropical countries and affecting millions of people. Adults live in the blood system, but it is the eggs which give rise to medical problems; eggs retained by the host cause pathological tissue reactions. Eggs not retained pass out in faeces or urine and if they reach water develop into a free-swimming larva, a miracidium. This must enter a second, intermediate host for further development. Intermediate hosts are certain species of aquatic snail, and in these a new lifestage develops, the sporocyst. A second generation of sporocysts may result but eventually sporocysts produce a second type of free-swimming larvae, the cercaria. Ultimately, a single miracidium may give rise to many thousand cercariae. Cercariae can penetrate the unbroken skin of man, and so infection occurs when skin contact is made with infected water. Overall, the probable absence of suitable intermediate hosts and the prevalence of good hygiene in Australia are likely to prevent the establishment of medically important schistosomiasis, but the matter needs watching, and certainly a careful guard must be maintained against the possible establishment of exotic snail hosts (there are occasional reports of these being imported in tropical aquarium fish consignments).

Of more significance in Australia is another platyhelminth parasite, *Fasciola hepatica*, the liver fluke of sheep and cattle. Its lifecycle has the general pattern described above, but the definitive hosts are sheep or cattle. Fascioliasis can cause severe economic losses and apparently some 25% of all sheep and cattle in Australia graze in areas at risk. The parasite was introduced to Australia but has been able to use a native aquatic snail, *Austropeplea* (formerly *Lymnaea*) *tomentosa*, as its intermediate host (Fig. 4.5(c)).

There are, of course, many platyhelminth parasites of native origin in Australia and infecting a variety of vertebrates associated with inland waters (e.g. waterbirds, snakes). Some may give rise to minor medical problems. Benign dermatitis or bather's itch is caused by the skin reaction set up when the cercariae of some of these species (mistakenly) attack bathers (Fig. 4.6).

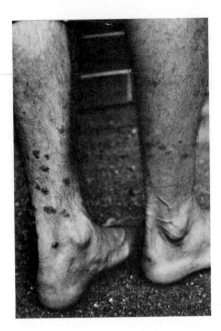

Fig. 4.6 Benign dermatitis. The photograph shows the skin reaction of a student after wading in Lake Alexandrina, South Australia, and being attacked by platyhelminth cercariae. Photograph by courtesy of the Institute of Medical and Veterinary Science, Adelaide.

Nematode worms, likewise, include many important pathogens whose transmittal between hosts implicates aquatic macroinvertebrates. Again, Australia is fortunate in that none is of any medical significance here, though significant in most other tropical countries including our near neighbour, Papua New Guinea. A variety of diseases is involved. Particularly important is elephantiasis (a gross swelling of body parts and a symptom of several sorts of infection, but especially by *Wuchereria bancrofti*), onchocerciasis (river blindness), and infection by the Guinea worm, *Dracunculus medinensis*. Life histories of the disease organisms display many patterns. *Wuchereria bancrofti* cycles between man and adult mosquitoes; it is mentioned because, of course, larval mosquitoes are aquatic in habit. The organism causing onchocerciasis cycles between man and blackflies (Simuliidae; Fig. 3.8(e)), another group of insects with aquatic larvae. And *D. medinensis* cycles between man and certain cyclopoid copepods (cf. Fig. 4.2(g)). Many millions of people are infected by all three of these diseases.

No other groups of aquatic invertebrates *per se* give rise to medically significant diseases. However, some aquatic insects (chiefly mosquitoes) are involved in two further sorts of disease that are or were important in Australia and that have not been mentioned so far. These are arbovirus diseases and malaria.

The term arbovirus is a generic one used broadly to refer to arthropod borne viruses. In Australia, many such viruses are present but only a small number are medically significant. They include a virus causing encephalitis (inflammation of the brain) formerly known as Murray Valley encephalitis virus (MVE) but now known as Australian Arbovirus, and one causing epidemic polyarthritis, the Ross River virus (RRV). Many species of mosquito are implicated as vectors of arboviruses (including species of *Culex, Aedes, Mansonia,* and *Anopheles*). It appears that man is only incidentally infected, and the natural host of arboviruses is a species of waterbird or terrestrial mammal (e.g. kangaroo). Nevertheless, arbovirus diseases are of some medical significance in Australia, though fortunately absolute incidence is not great.

Malaria, too, was formerly of medical significance in Australia. It has now been largely if not entirely eradicated. The actual pathogen is a protozoan parasite, *Plasmodium*. It cycles between man and mosquitoes. Though malaria is no longer significant here, a note of caution should be sounded. There is a constant threat of its reappearance, a threat posed by (a) the presence of the parasite in large reservoir populations in neighbouring countries, including Papua New Guinea, (b) the occurrence of mosquito vectors in Australia, (c) the rapidity of modern travel, (d) recent increases in the number of migrants from Asian countries where malaria is endemic, and (e) the great number of people now living in the north of Australia as a result of mining and other developments there.

Apart from a role as vectors of malaria and other diseases, several groups of aquatic insects, as indicated at the beginning of this section, can also represent significant pests. Here, it is usually the skin irritation from the bite of an aerial adult that is the reason why the insects are nuisances—although biting may be so intensive and reaction so severe that the effect may go well beyond nuisance value. In a few cases, it is sheer insect numbers rather than their bite that cause problems. Four families of Diptera are the principal insects involved: Simuliidae, Chironomidae, Ceratopogonidae and Culicidae. Of these only the ceratopogonids have some forms with non-aquatic larvae; the others, as well as the most pestiferous ceratopogonids, have larvae entirely aquatic in habit.

In the Simuliidae (Fig. 3.8(e)), blackflies, two species are important: *Austrosimulium pestilens* and *A. bancrofti*. The former is of considerable significance in Queensland where it may viciously attack both man and stock; *A. bancrofti* is a more widespread species. Adult gnats or midges

c

(Chironomidae) may be a nuisance not because they bite, but rather because they may be attracted to lights in numbers so massive as to cause considerable annoyance. In South Australia, the enormous numbers of *Polypedilum nubifer* attracted to lights near Bolivar sewage works (before control measures) and of a *Tanytarsus* species from salt lakes near Port Augusta, are two local examples known to me. The Ceratopogonidae (Fig. 4.4(g)), or biting midges, as their name suggests, often go beyond mere annoyance; many are notable pests and fiercely attack and bite both man and a number of other vertebrates. In southeastern Queensland where large numbers of *Culicoides molestus* and *C. subimmaculatus* breed in artificial canals in housing estates, midge densities are often so high that active control measures are needed to maintain habitable conditions. Ceratopogonids, it may be added, are also vectors of allergic dermatitis in horses. Reference has already been made to the Culicidae (Fig. 4.4(h)), mosquitoes, as disease vectors. Many other culicids are not important in this respect but are important nuisance insects, biting both man and other animals. *Culex fatigans* is an important species of this sort. Only female mosquitoes bite.

Finally, brief mention should be made of problems arising from aquatic invertebrates affecting animals other than man and stock. Fish parasitism is perhaps the most important of these. However, fish parasitology is an infant science in Australia so the extent of the problem remains to be determined. Until recently many exotic fish diseases appeared to be absent from Australia (with the somewhat anomalous result, for example, that Australia was an important source of disease-free trout eggs to the northern hemisphere!), but this situation is probably changing rapidly — a likely consequence of the profligate and unquarantined importation into Australia of millions of aquarium fish (see Chapter 5). An example of fish parasitism that has caused a problem is the recent introduction into Lake Burley Griffin, Canberra, of *Lernea*, a parasitic copepod. This has apparently affected trout populations in a most deleterious way. Leech attacks, the weakening of farm dam walls by freshwater crayfish and crab burrows, and damage to rice crops by certain chironomid larvae and notostracan crustaceans are some other, and usually minor, problems involving aquatic invertebrates.

USES

One obvious direct use is as food. Freshwater mussels and crayfish were certainly eaten by many aboriginal communities before the advent of Europeans to Australia, as numerous shell middens testify. In more recent times, a few species of freshwater crayfish have been exploited as food, though significant quantities are not involved. Certain crayfish species are currently the basis of aquacultural enterprises, especially the yabbie, *Cherax destructor*, in southeastern Australia, and the marron, *C. tenuimanus*, in Western Australia.

A less obvious practical use, but increasingly a more significant one, is to indicate water quality. With the rapidly deteriorating quality of many fresh waters it is certain that this use of aquatic invertebrates, viz as biological monitors, will come to assume much greater significance than at present. The values of biological monitoring as a means of assessment and monitoring water quality are basically five-fold: (a) a single set of samples sums (integrates) environmental conditions during the preceding period; (b) biological monitoring can detect intermittent pollution better than physicochemical methods; (c) likewise, biological monitoring can indicate the extremes of previous conditions better than physicochemical methods; (d) certain aquatic invertebrates are extremely sensitive to particular pollutants; and (e) biological monitoring provides an integrated picture of the effects of pollution by taking into account synergistic and antagonistic effects of pollutants on the natural environment in a way quite impossible for physicochemical techniques.

Biological monitoring also has a predictive use, and the usual form this takes with invertebrates is to use them in bioassays or toxicity tests. Basically, these involve the determination of given limits for pollutants in aquatic environments. It is early days for Australian investigations in this area, but already three aquatic invertebrates have been proposed as suitable test animals: *Austrochiltonia* spp. (an amphipod, Fig. 4.3(f)), *Paratya australiensis* (an atyid prawn, Fig. 4.3(g)) and *Velesunio ambiguus* (a freshwater mussel, Fig. 4.1(f)). The freshwater mussel also has potential as a particularly sensitive monitor of pesticides and heavy metals (though a number of difficulties are apparent).

Yet another practical use of aquatic invertebrates of increasing importance and potential is as biological control agents. Several species of mosquito already act in this way and transmit myxomatosis amongst Australian rabbits, so controlling population densities of this noxious animal. Of more potential than actual use at present is the use of sciomyzid flies to control aquatic snail populations. Nevertheless, there is considerable interest in this family of Diptera as possible agents for the biological control of many platyhelminth parasites. *Cypretta*, an ostracod genus found in Australia and elsewhere, may also prove to be a useful biological control agent of snails acting as intermediate hosts of parasites. A species of *Cypretta* has proved to be an effective predator in experiments conducted in the United States.

5   Fish

This chapter surveys the sorts of native (and introduced) fish in Australian inland waters, discusses their distribution and adaptations to the Australian environment, and considers various problems and other matters associated with introduced fish. Finally, there is a short discussion of some commercial matters.

Before proceeding further it should be stressed that we know very little about our native freshwater fish. Systematic knowledge is poor, and ecological knowledge is scanty. Indeed, we know more about the biology of introduced fish than native ones. The reasons are many and varied, but are of little concern here. It need only be said that much awaits discovery by Australian ichthyologists who perhaps may, in the words of a recent reviewer, bring about major changes in the complexion of the fauna.

THE SORTS OF FISH

Living vertebrates are usually divided into seven groups. Of these, fish or fish-like animals comprise three; the remaining ones are the amphibians, reptiles, birds and mammals. The first and most primitive of the 'fish' groups is the Agnatha, a group in which jaws are absent. The first vertebrates to evolve were in this group, but the only living examples are lampreys and hagfish (class Cyclostomata). The second group is the Chondrichthyes, fishes with jaws but with a cartilaginous, not bony, skeleton; the elasmobranch sharks, skates and rays belong here. The third group is the Osteichthyes, containing the coelacanths and lungfishes (class Choanichthyes), and all other bony fish (class Actinopterygii). Members of all three groups of fish occur in Australian inland waters (Table 5.1).

With regard to the detailed composition of the fish fauna (Table 5.1), three important general points should be made. First, its diversity is restricted; in all, less than 200 species spend their whole lives in fresh water. Additionally, about 50 can best be regarded as marine 'vagrants', another 20 or so migrate between fresh and marine waters, and about 20 or thereabouts have been in-

Table 5.1 Composition of the native fish fauna of Australian inland waters. Table derived from one given by McDowall (1981) with minor amendments from Lake (1978), McDowall (1980), and McDowall and Frankenberg (1981). The numbers of genera in the Ariidae, Plotosidae, Teraponidae, Eleotridae and Gobiidae are tentative.

Family	Common name	No. of genera*	No. of species*
Class Cyclostomata			
Geotriidae	Lampreys	1	1
Mordaciidae	Lampreys	1	2(2)
Class Chondrichthyes			
Pristiidae	Sawfishes	1	1
Class Choanichthyes			
Ceratodidae	Lung-fishes	1(1)	1(1)
Class Actinopterygii			
Anguillidae	Eels	1	4(1)
Clupeidae	Herrings	2(1)	3(1)
Scleropagidae	Barramundi	1	2(1)
Retropinnidae	Smelts	1	2(2)
Lepidogalaxiidae	Scaled galaxias	1(1)	1(1)
Galaxiidae	Southern trout	3(2)	20(18)
Prototroctidae	Southern grayling	1	1(1)
Aplochitonidae	Whitebait	1(1)	1(1)
Ariidae	Fork tail catfishes	1	2(1)
Plotosidae	Eel tail catfishes	4(2)	11(8)
Belonidae	Needle fishes	1	1(1)
Melanotaeniidae	Rainbow fishes	3(1)	13(11)
Atherinidae	Silversides, hardyheads	3(2)	11(10)
Synbranchidae	One-gilled eels	1	2(2)
Scorpaenidae	Scorpion fishes	1(1)	1(1)
Centropomidae	Glass fishes, silver barramundi	1	1
Ambassidae	Chanda perches, etc.	2(1)	7(4)
Percichthyidae	Australian basses, cods	3(3)	7(7)
Teraponidae	Terapon perches, grunters	8(3)	20(16)
Kuhliidae	Flagtails, mountain perches	3(2)	6(5)
Eleotridae	Gudgeons	15(4)	33(19)
Gobiidae	Gobies	8(3)	15(8)
Kurtidae	Nursery fishes	1	1
Soleidae	Soles	2	3(2)
Gadopsidae	Blackfish	1(1)	1(1)
Apogonidae	Cardinal fishes	1	1
Toxotidae	Archer fishes	1	2
Mugilidae	Mullets	1	1(1)
Bovichthyidae	Tupong, congolli	1(1)	1(1)

* Number of endemic genera or species shown in parentheses.

troduced. The latest enumeration available (see Table 5.1) provides a figure of 177 for 'freshwater' species (it should be remarked that this table is not to be regarded as definitive for it incorporates a certain degree of arbitrariness as to what constitutes a 'freshwater' species — many fish found in fresh waters spend part of their lives in estuaries or the sea). With further systematic investigation this figure will undoubtedly increase, but compared with other continental fish faunas will always be low (cf. Africa which has about 2000 freshwater fish species). The first general point to be made, then, is that the fish fauna is depauperate.

A second point is that many Australian fish have close evolutionary relationships with marine forms. This relationship has perhaps been somewhat overemphasized in the literature, for contemporary views are that the fauna is really a heterogeneous assemblage. Be that as it may, apart from three species, little of the Australian fish fauna can be regarded as comprising 'primary' freshwater fish, namely fish that evolved in fresh waters after the ancestral bonyfish had immigrated to that environment, that are now confined to fresh waters, and that cannot withstand marine immersion. Primary freshwater fish include lungfishes, scleropagids, most catfishes, and, notably, the carps. Of these Australia has but one lungfish (*Neoceratodus forsteri*) and two scleropagids (*Scleropages leichhardti, S. jardini*). The carps and their relatives, the Ostariophysi (Cypriniformes), which are so widespread in the northern temperate region, dominant in Asia, and common in Africa, are entirely absent from the Australian native fish fauna. Even the Australian catfishes (Ariidae and Plotosidae) are of secondary marine derivation; their ancestral stocks evolved first in fresh waters, returned to the sea, and then entered Australian fresh waters. Thus, many fish characteristic of fresh waters in other continents do not naturally occur here; salmonids, characins, cyprinids, perches and cichlids are all missing, to name but five of the major sorts of freshwater fish found outside Australia. In summary, five major elements are identifiable in the derivation of the fauna: old endemics (*Neoceratodus, Scleropages, Lepidogalaxias*), pantropical species (e.g. Ariidae), southern temperate forms (e.g. Galaxiidae), Indo-Pacific forms (e.g. the giant perch, *Lates*), and endemics of uncertain origin (e.g. many Percichthyidae).

Thirdly, although depauperate in terms of number of species, the fauna is highly endemic. Most Australian freshwater fish are found only in Australia (Table 5.1). The relatively few species that are not endemic are mostly shared with Papua New Guinea or with southern circumpolar landmasses. Thus, the Galaxiidae, one of the most diverse of all Australian fish families, has a few species occurring in Australia, New Zealand, South America and the tip of South Africa. The family is, however, unknown in the northern hemishpere.

I do not intend to attempt any summary of the major points concerning each of the families indicated in Table 5.1. Nonetheless, it would be remiss if at least some discussion of the more interesting native fishes were not provided. Those selected for discussion are the lampreys, lungfish, scleropagids, eels, southern trout, Australian grayling, the giant perch, Macquarie and golden perches, the Murray cod and its close relative the trout cod, and the blackfish. Many fisheries biologists would select a somewhat different group; the difficulty is that the Australian fish fauna is a highly interesting one despite its depauperate nature.

The primitive, eel-like lampreys are north and south temperate fish lacking scales, true teeth, jaws and paired fins. They are grouped into three families, one for northern hemisphere forms, two for southern forms. Typically, their lifecycle involves a parasitic adult which feeds in the sea, an ascent of rivers by non-feeding adults (velasia) which spawn in river headwaters, a small, blind, filter-feeding ammocoete larva which feeds in the mud and silt

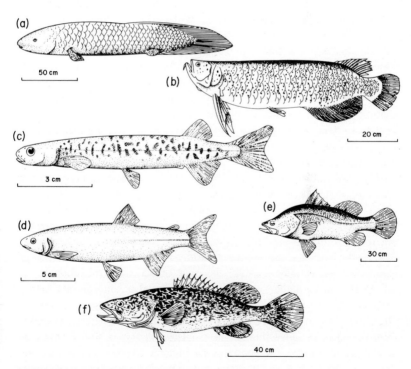

Fig. 5.1 (a) Lungfish; (b) spotted barramundi; (c) common galaxias; (d) Australian grayling; (e) silver barramundi; (f) Murray cod. Redrawn after various authors.

of rivers, a larval metamorphosis to an immature adult (macrophthalmia) with eyes, sucker and dorsal fins, and finally a seaward migration of macrophthalmia. Two of the Australian lampreys have this sort of lifecycle (*Geotria australis, Mordacia mordax*), but not the third (*M. praecox*). In *M. praecox*, the marine parasitic phase has been eliminated: the species is entirely non-parasitic and freshwater.

Neoceratodus forsteri, the Queensland lungfish (Fig. 5.1(a)), is yet another Australian freshwater fish belonging to an archaic group. It has survived relatively unchanged for more than 150 million years, and with *Protopterus* of tropical Africa and *Lepidosiren* of Amazonia, is one of only three extant lungfish genera. *Neoceratodus* is the least changed, but unlike its relatives cannot survive if its habitat dries. However, its lung does act as an accessory breathing organ when active. It presently occurs in several southeast Queensland rivers, but has been introduced into most of these, and the Burnett River is the only locality in which it occurs naturally with certainty. The Queensland lungfish is an omnivore with apparently some predilection for animal food.

Like the Queensland lungfish, the scleropagid (formerly, osteoglossid) fish of Australia represent remnants of an ancient and previously more widespread fish group. They are, however, not quite so archaic; though certainly old as bonyfish go, their ancestry is only some 50 million years. Confined to tropical South America and Africa, the Indonesian archipelago (including Papua New Guinea) and Australia, less than 10 living species are known. Two are Australian. The spotted barramundi (*Scleropages leichhardti*, Fig. 5.1(b)) is found in the Fitzroy River of Queensland, and the northern spotted barramundi (*S. jardini*) in several northern Australian rivers and Papua New Guinea. The feeding habits of the two species are somewhat different, though both are carnivores.

The freshwater eels, like the lampreys, migrate between freshwater and marine environments. However, unlike the *anadromous* lampreys, freshwater eels spawn in the sea and enter fresh waters to feed (they are *catadromous*). Also unlike lampreys, eels are not parasitic. The lifecycle of a freshwater eel involves spawning somewhere off the coast of Australia, migration of leptocephalous larvae to the Australian coastline, larval metamorphosis to elvers, upstream migration, growth to adults in fresh waters, and a downstream migration of adults when breeding time approaches. There are four Australian freshwater species, all within the genus *Anguilla*; this is absent only from southwestern streams. The nature of the lifecycle restricts eels to rivers and streams of reasonably easy coastal access, but eels do penetrate considerable distances inland, and overland travel during damp conditions is also possible. All eels are carnivorous.

A few galaxiids also migrate between fresh waters and the sea, but most are confined to fresh waters. The family, the Galaxiidae or southern trout, is regarded as perhaps the most characteristic of temperate Australian streams and lakes. It is entirely southern hemispheric in distribution, being found in Australia (20 species), South America (4), the tip of South Africa (1), and New Zealand (13). It also occurs on various islands in the southern hemisphere oceans: New Caledonia (1), Lord Howe (1), Chathams (4), Auckland and Campbell Islands (1), Falklands (2). Apparently Australia was the evolutionary centre of the family. The hypothesis attracting most support is that galaxiids evolved in the Australian region and dispersed eastwards by marine migration of salt-tolerant forms which subsequently invaded fresh waters. Another hypothesis is that present distributions result from continental drift. All but two Australian species are endemic, and most are local in distribution. A few, however, are widespread: *Galaxias maculatus* (Fig. 5.1(c)), for example, occurs throughout southeastern Australia as well as in New Zealand, South America and on various islands. This species is catadromous in typical environments, that is, it migrates seawards to breed—though the term catadromous should perhaps not be used since the species actually breeds in estuaries and it is from there that seagoing juveniles originate. Landlocked forms are also known; adults of these migrate *up* streams to spawn. *G. brevipennis* is another species with seagoing juveniles, but breeding appears to be in fresh waters alone. It too has landlocked populations undergoing no significant migration. Galaxiids have attracted a good deal of attention from zoogeographers, but seem zoogeographically overvalued in the light of the tolerance of some of them to seawater and the great distances—up to 700 km—that specimens have been recorded away from land.

The Australian grayling, *Prototroctes maraena* (Fig. 5.1(d)), was regarded until very recently as one of Australia's rarest fish and the most endangered freshwater fish species. Fortunately, observations over the past few years suggest that this may not be the case: it is a good deal more common than previously thought. Whatever the situation, it is only recent observations that have provided any substantive information on the biology of the species. Indeed, so little was known about the species that it had been placed in three different families by ichthyologists before a new family, the Prototroctidae, was created for it. There is a single Australian species in the family, and its only congener, a New Zealand species, is now thought extinct. *P. maraena* occurs in coastal rivers of southeastern Australia (including Tasmania). Present evidence is that the breeding season is short and synchronized. Spawning is in autumn in freshwater reaches of coastal rivers, with fry subsequently being swept downstream to brackish areas. Juveniles migrate back to fresh waters in spring. Adults live in fresh waters, school, and are highly fecund omnivores.

One species of the Centropomidae deserves mention here: the giant perch, silver barramundi, or, simply and most often, the barramundi (*Lates calcarifer*, Fig. 5.1(e)). This species is widely distributed throughout the warmer parts of the Indo-Pacific region including many rivers in northern Australia. It is perhaps the best known northern freshwater fish, being of considerable commercial and sporting significance; it grows to a size of almost 2 metres and 60 kg. Individuals are spread by monsoonal floodwaters to isolated billabongs and river pools where growth, but no breeding takes place. To breed, adults must migrate to estuaries. Because subsequent return may be prevented by impoundments, many river populations face extinction.

In contrast to the Centropomidae, where only one species is of any significance as a sporting or food fish, almost all Australian species of Percichthyidae are significant: the family includes the largest Australian freshwater fish, the Murray cod (*Maccullochella peeli*, Fig. 5.1(f)), and some of the best native sporting and food fish. The Murray cod may reach weights in excess of 100 kg, though 30 kg is now a more typical maximum. Not surprisingly for so large a fish, it is a carnivore, and birds, rodents and even platypus figure in its diet. It occurs throughout the Murray–Darling basin as well as in southeastern Queensland and northwestern New South Wales. Until recently it was confused as a single species with its much smaller relative, the trout cod (*M. macquariensis*), the scientific name of which was formerly (and is now incorrectly) applied to the Murray cod. The trout cod survives in only a few isolated localities. The Macquarie perch (*Macquaria australasica*), though now less widely spread than formerly, is found in the upper reaches of the Murray–Darling system and in some coastal rivers of New South Wales. Like the Murray cod, it is a significant native sport fish, but its importance in this respect is declining. Attempts to induce spawning artifically have been largely unsuccessful. Artificial spawning, however, has successfully been induced in another important percichthyid, the golden perch (alias callop, yellowbelly, Murray perch, white perch or *Macquaria ambigua*). This species occurs throughout the Murray–Darling system as well as in other inland and in coastal river systems; it is regarded as our most widespread and edible native fish. Murray cod may also be induced to spawn artificially.

Finally, mention should be made of the blackfish. At present only one species, *Gadopsis marmoratus* (Fig. 5.2(a)), is formally recognized from both the mainland of southeastern Australia and Tasmania, but Tasmanian populations have been informally described as a separate form. On the mainland, *G. marmoratus* occurs in the upper reaches of the River Murray and in rivers and streams of Victoria. Since the species is the only one within the family, the Gadopsidae constitutes an Australian endemic fish family. It has an interesting combination of advanced and primitive anatomical characters, and has been

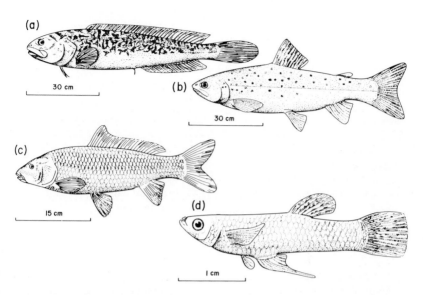

Fig. 5.2 (a) Blackfish; (b) brown trout; (c) common carp; (d) mosquito fish (male). Redrawn after McDowall (1980).

regarded by some as a primary freshwater fish family. If so, it would seem to have evolved entirely in fresh waters within Australia. Whilst less abundant than before, the mainland form is still relatively common. It has adhesive eggs which are deposited inside hollow logs. The diet consists of both aquatic (mainly) and terrestrial insects.

Thus far, discussion has been confined to native fish. However, many freshwater fish have been introduced either deliberately or unintentionally (Table 5.2), and, though of less interest biologically, they have had an impact upon the freshwater environment and merit discussion.

The salmonid introductions have unquestionably proven the most popular, and five species have been introduced of which at least three maintain self-reproducing populations. The first and by far the most successful introduction was in 1864 when eggs of brown trout (*Salmo trutta*, Fig. 5.2(b)) were successfully transhipped from England to Tasmania. This attempt, however, had been preceded by several unsuccessful ones, beginning first, it seems, in 1841. Success came in 1864 when 90 000 salmon eggs and 1500 brown trout eggs (together with about 30 tons of ice) were shipped aboard the *Norfolk* and left Falmouth, England, on 28 January 1864. They arrived Melbourne on 15 April 1864. Eggs of both species survived the journey, and after transhipment to Hobart and thence Plenty (on the River Plenty, a tributary of the Derwent), fry hatched in May 1864. The first releases occurred in 1866; of these the trout

Table 5.2　Introduced freshwater fish. Table compiled from Tilzey (1980), Cadwallader, Backhouse and Fallu (1980) and Trendall and Johnson (1981), but incorporating some unpublished information of B. J. McMahon (Queensland Institute of Technology) and others.

Scientific name	Common name
Salmonidae	
Salmo trutta	Brown trout
S. gairdneri	Rainbow trout
S. salar *	Atlantic salmon
Oncorhynchus tschawytscha *	Quinnat salmon
Salvelinus fontinalis	Brook trout
Poeciliidae	
Gambusia affinis	Mosquito fish
G. dominicensis	Mosquito fish
Poecilia reticulata	Guppy
P. latipinna	Sailfin molly
Xiphophorus helleri	Swordtail
X. maculatus	Platy
Phalloceros caudimaculatus	—
Synbranchidae	
Fluta alba†	Belut
Amphipnoidae	
Amphipnous cuchia†	Cuchia
Percidae	
Perca fluviatilis	Redfin
Cyprinidae	
Cyprinus carpio	Common carp
Carassius auratus	Goldfish
C. carassius (?)	Crucian carp
Tinca tinca	Tench
Rutilus rutilus	Roach
Puntius conchonius	Rosy barb
Cichlidae	
Tilapia mariae	Black mangrove cichlid
Cichlasoma nigrofasciatum	Convict cichlid
C. octofasciatum	Jack Dempsey cichlid
Seratheradon mossambica (?)	Mozambique mouthbrooder

* No evidence of natural reproduction.
† Probably or almost certainly introduced.

survived, but not the salmon. A second viable consignment arrived from England in 1865 on board the *Lincolnshire*. From Plenty, brown trout were taken to Victoria and thence, in 1889, to New South Wales. Introductions to Western Australia were unsuccessful until 1931. The species now survives in nearly all suitable localities in southeastern Australia including Tasmania where it is absent from only the most inaccessible waters of the southwest. It also occurs, but does less well, in South Australia and Western Australia.

North American rainbow trout (*S. gairdneri*) first reached Australia, via New Zealand, in 1884. Slightly hardier than brown trout, it is now as widespread, though slight differences in certain responses have been demonstrated between geographical strains. Both this species and the brown trout survive throughout most of their range of distribution in Australia by natural spawning. They also occur in many localities (e.g. farm dams) where suitable spawning facilities are absent; here they survive by continual restocking with material artificially bred in hatcheries. The key factors responsible for the success in the establishment of *S. trutta* and *S. gairdneri* have been suggested as (a) the environmental similarities between Australian and ancestral habitats, (b) minimal competition from indigenous fish, (c) the abundance and availability of food, and (d) the virtual absence of parasites and disease.

The quinnat salmon (*Oncorhynchus tschawytscha*), 'successfully' introduced in 1936 to various Victorian lakes, survives entirely as a result of artificial stocking. However, a population recently introduced into New South Wales may be self-sustaining. Earlier attempts to introduce this species to Victoria (1877) and Tasmania (1910) were unsuccessful. The brook trout (*Salvelinus fontinalis*) and Atlantic salmon (*Salmo salar*) represent recent salmonid introductions, the success of which is uncertain. The brook trout, a native of North America, has been introduced into New South Wales and Tasmania, and the Atlantic salmon into New South Wales where, its protagonists hope, it will maintain itself as a landlocked form of this species. Early introductions of salmon to Tasmania, as noted, were unsuccessful.

Cyprinid introductions have also been important, though far less popular than salmonid introductions. At least five species are involved. Tench (*Tinca tinca*) was introduced to Tasmania and then Victoria and New South Wales in the late nineteenth century. It now inhabits the lower reaches of rivers and standing waters associated with the Murray–Darling system as well as similar habitats in Tasmania. The roach (*Rutilus rutilus*) has a more restricted distribution; it is found only in some southern Victorian rivers and rarely in the Murray. Of the two species of *Carassius* listed in Table 5.2, the occurrence of one, *C. carassius*, has been reported but not confirmed, whilst the other, *C. auratus*, the goldfish, is now widespread in all states wherever suitable sluggish and standing waters occur. It was introduced about 1876.

The most significant cyprinid introduction has been that of the common, Prussian, or European carp (*Cyprinus carpio*, Fig 5.2(c)). The question of the environmental impact of this species will be taken up later; here only some brief notes on its history and distribution in Australia need be given. It was first introduced about 1865 near Sydney. A second introduction occurred at some time prior to 1903 in the Murrumbidgee irrigation area. Neither introduction was followed by a rapid spread. Then, about 1960, and apparently after the illegal importation of a fresh strain, this species underwent a population explosion and began to spread rapidly. Initially the spread was a Victorian affair and the Victorians made every effort to curb it: emergency legislation went through the Victorian parliament in May 1962 declaring the fish a noxious species, and the Victorian Fisheries and Wildlife Department, as it was then, made determined efforts to eradicate every known population of the fresh strain. They lost the battle. As a result the European carp now infests most of the lower part of the Murray–Darling system and many parts of Victoria, New South Wales and Queensland. It has, at the time of writing, penetrated the Murray upstream to Albury. It does not yet occur in Western Australia, but it is highly tolerant to many environmental stresses (including elevated salinities) and it is reasonable to predict that one day it will be distributed throughout Australia.

The European or English perch (*Perca fluviatilis*) — known in Australia as redfin — is yet another important introduction. It was introduced first to Tasmania (1862), then Victoria (1868) and New South Wales (1888). It now also occurs in South Australia and southwestern Western Australia. In the lower Murray system, it is so common that a small commercial fishery is supported.

None of the remaining introduced species (Table 5.2) has any piscatorial significance, and apparently only the mosquito fish has any biological importance: the belut and cuchia are rare, and the rosy barb, sailfin molly, guppy, swordtail, platy and cichlids are represented by isolated feral populations derived from aquarium fish released in eastern Queensland or Victoria. *Phalloceros caudimaculatus* has a feral population in Western Australia. The mosquito fish (*Gambusia affinis*, Fig. 5.2(d)) was introduced earlier this century to control mosquito larvae. There is no evidence, however, that it does so better than native fish. It is now widespread on the mainland, including southwestern Western Australia, South Australia, and even semi-arid areas of central Australia. This small fish is an extremely hardy and voracious species that has probably had considerable impact upon the native freshwater fauna. Since it is of little significance to anglers, there has been little research on this subject.

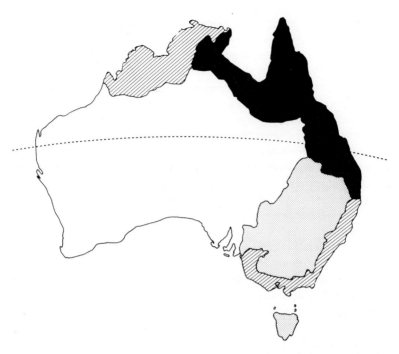

Fig. 5.3 Distribution of freshwater fish in Australia according to major drainage areas and number of species in each area. ■ 50–59 species; ░ 40–49 species; ▒ 20–29 species; ☐ 10–19 species.

DISTRIBUTION

Figure 5.3 indicates the distribution of native and essentially freshwater fish in Australia according to major drainage areas and the number of species in each area. Clearly, over much of Australia, few species occur, a fact not surprising in view of the aridity of much of the continent. However, even in moist temperate areas such as Tasmania and southwestern Western Australia there are few species, and in the vast Murray–Darling drainage area (over 1 million km²) only 27 species occur. Diversity is greatest in the tropical north and northeast where, on the other hand, endemicity is least; here is found the greatest number of species also found outside Australia (some two-thirds of the species also occur in Papua New Guinea).

History, ecology and chance all play a part in determining Australian native freshwater fish distributions, but no comprehensive analysis of the fauna in these terms has been attempted (or, indeed, is possible given the present poverty of our knowledge). Some general geographical relationships,

however, are obvious. Grayling, smelts, galaxiids and blackfish are mainly
southern temperate in distribution, whereas several other families are mostly
tropical (scleropagids, rainbow fish, eel tail catfishes, grunters, silversides, cen-
tropomids). Considering regional distributions, the Tasmanian fish fauna is
dominated by galaxiids with migratory species not unimportant. In the
southeastern coastal area, galaxiids and migratory fish remain important, but
several more species of goby occur in addition to other families. The Murray–Darling area has fewer families and species represented, but the perich-
thyids or basses are an important group. In northern and northeastern areas,
six families are important: eel tail catfishes, rainbow fish, silversides, centro-
pomids, grunters, and gobies. Three of these are also important in the fish
fauna of more arid areas: gobies, grunters and eel tail catfishes. Finally, in the
temperate southwestern coastal area, galaxiids and gobies are the most
important.

Quite how fish disperse within and between areas remains unknown for
most species. No doubt a large proportion of the fish fauna is more or less
passively spread by floodwaters. Dispersal of this sort seems to be a typical
part of the lifecycle of many northern fish. Dispersal in water spouts may also
occur, although its significance is uncertain. Hydrographical changes in river
systems, such as the capture of headwater streams, represent another possible
method of dispersal. Temporary marine migration between river mouths is
another, and even the spread of introduced trout seems sometimes to have involved such migrations. A few fish can disperse directly overland; eels may do
this, as well as the unrelated one-gilled eels (Synbranchidae) which can breathe
air. Finally, note should be made of the influence of man in fish dispersal, for
apart from an obvious influence in the spread of introduced fish, several native
fish distributions have been changed as a result of transference of species from
one river system to another.

The sorts of inland water bodies in which fish are distributed are
remarkably varied. Apart from such obvious environments as permanent
freshwater lakes and rivers, fish have been recorded from saline lakes, cave
pools, artesian wells, desert waterholes, isolated mound springs, and large
temporary lakes. Not all of these environments can be discussed here.

In the saline lakes of southeastern Australia, one species, *Atherinosoma
microstoma*, has been recorded at salinities up to 100°/∘∘, a value almost three
times that of sea water. However, essentially, this species occurs in coastal
lakes and lagoons, and the maximum salinity at which truly inland fish occur is
considerably lower; the three inland species most tolerant to salinity appear to
be the Lake Eyre hardyhead, the mosquito fish and the common galaxiid.
These occur at salinities up to about 30°/∘∘. They also occur in fresh waters.
There appear to be no truly halobiontic fish species in Australia, that is, fish
confined to inland salt waters.

Records of fish from caves and wells in Australia are few, but there are some. They include *Milyeringa veritas* (Eleotridae) and *Anommatophasma candidum* (Synbranchidae) from a well in Western Australia. Records of fish in desert waterholes and other isolated waters are more numerous. Special reference is made in this connection to *Chlamydogobius eremius*, a small goby confined to central Australian waters.

Before leaving the subject of fish distributions, a note should be made of the great distances travelled upstream by many fish of essentially marine habitat. As indicated previously, about 50 such species (in some 20 families) have been recorded from Australian fresh waters; many of these records were well away from the coast.

ADAPTATIONS

Adaptations of fish to life in inland waters are many and varied. Of special interest here are those to cope with the unusual nature of Australian waters, above all, their unpredictability. Breeding adaptations are amongst the more important. Several species have reproductive cycles synchronized to hydrological regimes so that spawning occurs only after a rise in water level: breeding follows flooding. In this way, juvenile fish develop where abundant food is available because flooding also stimulates the development of other aquatic organisms. Fish exhibiting this pattern of breeding are notably the golden perch and the silver perch (*Bidyanus bidyanus*). In other species, the spangled perch (*Leiopotherapon unicolor*) for example, flooding helps to induce breeding, although is not an essential factor. Yet others do not require a flood to stimulate spawning, but nevertheless find flood conditions advantageous: larger recruitment to populations of Murray cod, for example, is reported in flood years than in non-flood years. Species which require flooding to breed resorb eggs in the absence of suitable conditions.

Another reproductive adaptation apparently related to the nature of Australian fresh waters involves the type and number of eggs produced and the nature of subsequent parental care (if any). Large numbers of floating (pelagic) eggs are produced by the golden perch, a most unusual phenomenon for freshwater fish; freshwater fish typically have eggs heavier than water (demersal eggs). For a fish inhabiting a river system which periodically inundates vast areas of low-lying floodplain, as does the River Murray wherein the golden perch is found, the occurrence of floating eggs is by no means disadvantageous. The silver perch, too, produces free-floating eggs, although these require some water movement to avoid sinking. The eggs of this species, like those of the golden perch, hatch quickly and have large, peripheral 'cushions' (perivitelline spaces) to protect them against jarring: clearly, two further adap-

tations to the unpredictability of the environment inhabited by the species. In
other species, eggs are demersal, but may be adhesive or non-adhesive, laid in
nests or scattered, have parental care or not, and be numerous or few. Whilst
these features may not directly relate to Australian conditions in a unique
sense, it has been suggested that the very diversity of reproductive mechanisms
may itself be an adaptation; that is to say, in an unpredictable environment,
there has been selection for reproductive diversity. The most elaborate
methods of parental care are displayed by those species that are buccal
breeders, those incubating eggs orally. Females of the spotted barramundi
(*Scleropages leichhardti*) do this, as do males of the lesser salmon catfish (*Hex-
anematichthys leptaspis*) and mouth almighty (*Glossamia aprion*). A more bizarre example of parental care is provided by the nursery fish (*Kurtus gulliveri*);
the eggs are carried by the male on the end of a hook from his forehead.

Other adaptations, apart from reproductive ones, that relate to the nature
of Australian inland waters include the ability of some fish to aestivate (withstand prolonged drying) and the physiological tolerance of several to wide
variation in environmental conditions. Authenticated examples of Australian
fish that aestivate are rarer than might be expected. However, there is no
doubt that some species of galaxiid burrow in damp mud or earth to survive
dry periods in temporary streams. The spangled perch may also do so, though
the evidence is less firm. It has been claimed that Aborigines used to collect the
species from dried mud, but firm experimental evidence is lacking. Moreover,
although the species has been collected from ephemeral surface waters — only
circumstantial evidence of aestivation — so also have others not known to
aestivate (e.g. *Chlamydogobius eremius*). Floodwaters could have carried fish
to these localities.

There is no question, on the other hand, that the spangled perch is a fish
having a wide physiological tolerance to extremes of temperature. The lower
lethal temperature for it is about 5°C and the upper one almost 40°C.
Chlamydogobius eremius, likewise, is able to tolerate a wide range of
temperature (9–40°C). Additionally, it can survive in salinities between 0.2
and 37°/oo, and in oxygen levels less than 1 ppm (although at low oxygen concentrations aerial respiration may occur). The wide physiological tolerances of
this fish are regarded as the principal adaptation to the tenuous, fluctuating,
and unstable environment it inhabits, the isolated waters of the Lake Eyre
drainage basin. There is no evidence for aestivation in this species, nor, in fact,
in any central Australian fish. Various grunters (*Scortum* spp.) are also extremely hardy and can survive temperatures up to 40°C. A few fish, apart
from perhaps *Chlamydogobius eremius*, are able to tolerate low oxygen levels
by respiring atmospheric oxygen when necessary. The Queensland lungfish can
do so, as well as when extra oxygen is required for increased muscular activity.

Finally, two adaptations not related specifically to the nature of Australian fresh waters but of sufficient interest to merit comment may be noted. The archer fish, *Toxotes chatareus*, of northern Australian waters, has the interesting ability to spit water a considerable distance to catch prey. And Cox's gudgeon (*Gobiomorphus coxii*), a small goby, has its ventral fin modified to form a sucker so efficient that even vertical surfaces in running waters may be climbed. It is obviously well adapted to life in streams and rivers.

INTRODUCED FISH: SOME PROBLEMS

Opinions are divided about the impact of introduced fish upon native fish. One school maintains that the impact has been slight: altered habitats, changed hydrological regimes, and pollution of various sorts have been more important than any interaction between introduced and native species. The other school maintains that introduced fish have had considerable impact.

There is no doubt, of course, that man's modifications to the environment have disadvantaged some native fish, and that some introduced fish have scarcely affected native fish. But there are indications that that is not the whole story. There is evidence that trout on the mainland have replaced galaxiids in the middle reaches of some streams, with galaxiids persisting only in upper reaches inaccessible to or unsuitable for trout, or in lowland reaches with abundant weed cover. There is also some evidence that trout have affected the distribution of some freshwater invertebrates. The Tasmanian mountain shrimp (*Anaspides tasmaniae*) seems to survive only where trout predation is not heavy or where trout are absent.

The impact of other introduced fish is still less certain, and this will probably always be the case. Indeed, no complete assessment of the impact of even recent introductions has yet been made. The impact of the common carp, whose populations have exploded in southeastern Australia in the past two decades, still remains indeterminate. There is circumstantial evidence that it seriously damages the aquatic habitat by its habit of roiling, that is, by disturbing bottom sediments to find food. This makes the water turbid and may dislodge submerged plants. Its vigorous spawning habits may have the same impact. Additional potential impacts are on the feeding of native fish, on the natural distributions of certain waterfowl species, and on the composition of aquatic invertebrate communities. But firm unequivocal evidence on the reality of all these impacts in Australia has yet to be presented.

The nature of impact of the mosquito fish, likewise, remains uncertain. Nevertheless, for this fish, too, there is increasing evidence of a significant and negative impact. The species is widespread, hardy, and can tolerate both high salinities and temperatures. Voracious, fecund, and giving birth to live young, it would be remarkable if it had *not* had an impact on Australian freshwater fish and some invertebrates.

Notwithstanding our lack of knowledge about the impact of introduced fish already here, pressure continues to introduce still more! Often, it seems, this pressure would like to ignore the sort of procedure that concerned ichthyologists and freshwater biologists would regard as mandatory before the introduction of any fish (as outlined, for example, by Courteney and Robins (1973)). It also often seems to ignore the fact that general environmental disadvantages associated with an introduction may outweigh particular sectional advantages to proponents of an introduction.

One species recently proposed for introduction is the grass carp (*Ctenopharyngodon idella*). The proposal is based on the fact that this species eats nuisance aquatic plants. However, unfavourable side-effects may also eventuate, and these, too, should be considered. Another proposal involves the Nile perch, *Lates niloticus*. The rationale for its proposal is basically two-fold: (a) populations of the barramundi, *Lates calcarifer*, the only native fish of angling significance in northern Australia, are declining due largely to river impoundment; (b) the best way of compensating for this is to introduce from Africa a related species, viz the Nile perch, which is a good angling species and has a lifecycle that would not be disrupted by river impoundment. Strong arguments have been advanced against the proposal. In brief, they are that: (a) the Nile perch may compete with and affect remaining populations of the barramundi as well as those of other native fish (and invertebrates), (b) the introduction may not be necessary given the likely ability to induce artificial spawning in the barramundi.

Firm standards must continue to be maintained if proposals such as the above are not to succeed without proper safeguards. If standards are not maintained, then it seems only a matter of time before Australia becomes the cesspool of introduced fish species that Florida and other parts of North America are said to have become, and Papua New Guinea is well on the way to becoming. Special note should be made here of the widespread occurrence in Papua New Guinea of the Mozambique mouthbrooder (*Seratheradon mossambica*). Already there are reports of its presence in Queensland. It would be particularly unfortunate if this species were to become established in Australia.

Yet further dangers of this sort are posed by the importation, breeding and transport of aquarium or ornamental fish. The aquarium fish trade in Australia is now a million dollar industry involving several million individual fish per year. Over 10 million live fish are imported annually representing a value in excess of $1 000 000. Moreover, interest in ornamental fish is growing rapidly so that future values are likely to exceed considerably present ones (in 1963-64, the number of live fish imported to Australia was only 278 000 with a value of $21 000). The current world retail value of the ornamental fish trade exceeds $3 000 000 *per annum*. Most fish involved come from tropical and subtropical countries, but Singapore, Hong Kong and Thailand are important sources for Australia. South America is an important world source; Australia itself is not important at present, though several suitable species occur in northern waters. Such a vast trade in fish is not without dangers, and these are exacerbated by the less rigorous control accorded aquarium fish as compared with non-ornamental fish. The principal dangers are the possibilities of establishment of viable feral populations by undesirable aquarium species, and the possibilities of adding to the number of fish diseases already present in Australia. The possibility also exists of introducing diseases other than those of fish: we should remember that live imported fish bring with them non-sterilized imported water. Danger lies in the disposal of this water: a number of human pathogens is transmitted by water, and a number of disease vectors is aquatic. Although these possibilities have been brought to the attention of governmental authorities, most freshwater biologists would regard present legislation as inadequate despite some recent changes to it.

COMMERCIAL MATTERS

Freshwater fish provide a living for many Australians, and at least some brief discussion of the more important commercial features of freshwater fisheries should be included in the present chapter. Discussion will concern fish farming, fish hatcheries and commercial fishing.

The cultivation of fish as a crop in enclosed water-bodies and under carefully controlled conditions, intensive aquaculture, is an activity going back more than 3000 years. Apparently practised first by the Chinese, it has been a central European activity since at least medieval times. Fish are now farmed in this way widely, particularly in Europe and the Middle and Far East. In the East, valuable supplementary protein to the diets of indigenents is provided. The practice is gradually spreading in southeastern Australia as a commercially viable enterprise, but not, as in the East, to provide needed protein; in Australia, the main aim is to provide gourmet food. Rainbow and brown trout are the most popular species. There are few suitable native fish, for not many can breed in artificial ponds. Carp, the usual fish cultivated in many temperate regions, is not farmed in Australia, though illegal attempts to do so occurred in the late 1960s. Of course, there are many farm dams growing edible fish on a scale other than commercial, and involving a variety of fish. Trout are favoured, but native fish sometimes figure: the freshwater catfish, Murray cod, and silver and golden perch are four species suitable for southeastern dams. About the same number are suitable for northern, warmer waters. Cultivation of certain crayfish species in dams also elicits considerable interest. Extensive aquaculture, the cultivation of fish under less carefully controlled conditions and usually in natural water-bodies, also takes place in Australia, but not widely. One enterprise known to the author involves the transfer of large numbers of young eels caught in estuaries to slightly saline lakes in western Victoria. The enterprise is commercially viable and most of the product is exported.

Related to the subject of fish farming is that of artificial hatching of fish. Previously, this was practised widely in Australia, particularly with trout, the object being to release large numbers of young fish into the wild for the benefit of anglers. The rationale was simple: the more released, the bigger the final catch. Nature, however, does not function quite so simplistically, and we now know that profligate release of hatchery-bred individuals is useless. Hatcheries, of course, serve a useful function when natural breeding is impossible, as is the case for quinnat salmon in certain Victorian lakes. By and large, however, the trend in modern fisheries' practice is to discontinue artificial hatching of fish for release to the wild except under special circumstances. Most

Table 5.3 Freshwater fish of commercial importance in Australia. Data derived from Australian Bureau of Statistics.

Species	Major catch (state)*						
	Tas.	Vic.	NSW	SA	Qld	WA	NT
Short-finned eel	2	1					
Bony bream				1			
Tasmanian whitebait	1						
Freshwater catfish			2	1			
Silver barramundi					2	3	1
Murray cod			2	1			
Golden perch			2	1			
Australian bass		1					
Macquarie perch		1					
Silver perch			1	2			
Common carp	1	2	3				
Redfin	1	2	3				
Tench	2	3	1				
Rainbow trout				1			

* Numbers indicate order of importance.

hatchery-bred fish now go to artificial water-bodies, and, in fact, there is a small but nonetheless vigorous trade in the production of fry of both trout and several species of native fish for the stocking of farm dams.

Commercial catches of Australian freshwater fish for consumption are not significant compared with marine catches (and even these are insignificant on a world basis). However, that is not to say that there is none. On average, around 1000 tonnes live weight of freshwater fish are caught annually, representing some 1–2% of total Australian fish production (precise figures are difficult to obtain because data are not always separated into those applying to marine and those to freshwater catches). The principal species involved, together with an indication of where important, are listed in Table 5.3.

6 Amphibians and reptiles

AMPHIBIA

Of the three extant subclasses of amphibians, only one, the Anura (frogs and toads), is native to Australia. Urodeles (salamanders and newts) and caecilians do not occur here naturally. Although a great number of adult Australian frogs occurs considerable distances from free-standing water, there are many that are more closely associated with inland waters, and most (but not all) have an aquatic larval stage (tadpole) and therefore require water for breeding. A discussion of frog biology obviously has a place in a book of this sort.

All native Australian frogs have been placed in four families: Hylidae, Leptodactylidae, Microhylidae and Ranidae. A fifth family, the Bufonidae, includes an introduced species. There is some debate at present on the familial classification of Australian frogs, but to avoid confusion the older conservative classification is adhered to. A complete list of genera recorded, together with an indication of species diversity and general distribution patterns, is provided in Table 6.1. Readers are warned that whilst this table represents the most up-to-date enumeration at the time of writing, it is likely to be an underestimate of species numbers by the time of publication; descriptions of new species are appearing constantly as the result of increased intensity of investigation in the northern regions. It is now abundantly clear that the diversity of the Amphibia of northern Australia, particularly northern Queensland, is much greater than was previously thought. Representatives of the four Australian frog families are illustrated in Figs 6.1 and 6.2.

Most adults of the Hylidae (Fig. 6.1(a)–(e)) live in trees and low-growing shrubs, hence the common name 'tree-frogs', but at least some species are more dependent on watery habitats. *Litoria aurea*, for example, a common southeastern species, is usually found amongst vegetation within or bordering free water. An even more aquatic species is *L. nannotis* (Fig. 6.1(a)), a species restricted to northeastern Queensland; it is found in rocky streams and when disturbed swims beneath the water surface and hides under submerged rocks. Species of Hylidae occur throughout Australia, though some are rarely seen. Many have strong adaptations to arboreal life which need no description here.

Table 6.1 Systematic résumé of native Australian frog fauna, together with an indication of generic distribution. This table was prepared by M. Davies and M. J. Tyler, July 1981.

Family	Genus	Distribution	No. of species
Hylidae	*Cyclorana*	Northern half of continent	13
	Litoria	Widespread	53
	Nyctimystes	Northeast Qld	3
Leptodactylidae	*Adelotus*	Eastern coastal	1
	Assa	Southeastern Qld	1
	Crinia	Southeast, southwest	2
	Ranidella	Widespread	13
	Geocrinia	Southeast and southwest	4
	Kyarranus	Southeast	3
	Philoria	Mt Baw Baw, Vic.	1
	Taudactylus	Southeast Qld	5
	Pseudophryne	Widespread southern	12
	Arenophryne	West-coast, WA	1
	Limnodynastes	Widespread	12
	Lechriodus	Coastal Qld, northern NSW	1
	Uperoleia	Northern and eastern	16
	Rheobatrachus	Southeast Qld	1
	Myobatrachus	Extreme southwest	1
	Neobatrachus	Southern, coastal, interior	7
	Heleioporus	Central east coast southwards, extreme southwest	6
	Mixophyes	Central and northern east coast	3
	Notaden	Northern interior	3
	Megistolotis	Northwest	1
Microhylidae	*Sphenophryne*	Northeast Qld, NT	3
	Cophixalus	Northeast Qld	5
Ranidae	*Rana*	Northeast Qld	1

The genus *Litoria* (formerly *Hyla*) is the most diverse of all Australian frog genera: over 50 species have been described thus far.

The Leptodactylidae (Figs 6.1(f), 6.2(a)–(c)) comprise a rather heterogeneous and diverse group of frogs found throughout Australia. Adults range in habit from those that burrow to those that appear to spend a considerable time within water. Of the latter, *Rheobatrachus silus* (Fig. 6.2(c)) is the most aquatic. It occurs in streams not far from Brisbane yet was only recently discovered (Liem 1973). Well-adapted for life in water, it has deeply-webbed feet, is an adept swimmer, and has the ability to remain submerged for long periods at a time; it can remain underwater for several hours.

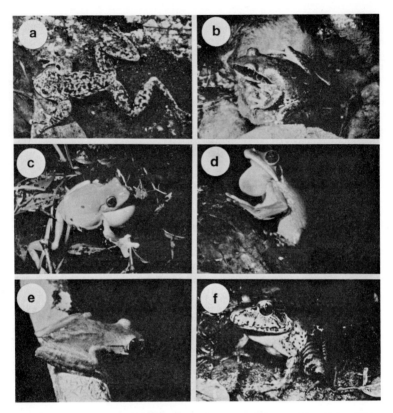

Fig. 6.1 Frogs. Representatives of the families Hylidae and Leptodactylidae. (a) *Litoria nannotis* from a stream in northern Queensland; (b) *Litoria lesuerii*, a widespread frog of eastern Australia, typically near streams; (c) a male *Litoria chloris* calling in a flooded pond near Erimba, NSW; (d) *Litoria infrafrenata* male calling near Daintree, Queensland; (e) *Nyctimystes tympanocryptis* from the Mulgrave River, north Queensland; (f) *Mixophyes iteratus* from an east coast rainforest. All photographs by courtesy of Dr Gordon Grigg, University of Sydney.

Strictly, perhaps, the Microhylidae (Fig. 6.2(d)) should not be considered, for all Australian species are terrestrial both as adults and larvae. They are mentioned merely to complete the picture, and require no more than a note that less than a dozen species are recorded from northern Australia.

The Ranidae is even less diverse than the Microhylidae; there is only one Australian representative, *Rana papua* (*R. daemelii*) (Fig. 6.2(e)). The family is basically northern hemispheric in zoogeographical distribution, and intrudes into Australia only in the northeast where *R. papua* is apparently fairly com-

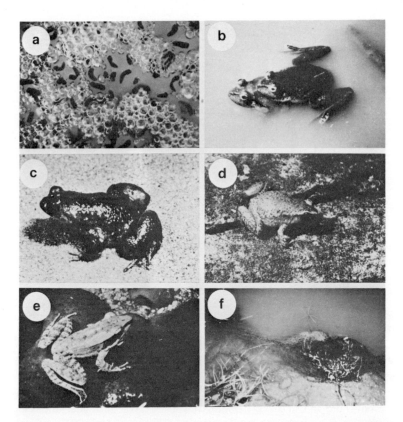

Fig. 6.2 Frogs. Representative lifestages of the families Leptodactylidae, Microhylidae, Ranidae and Bufonidae. (a) *Lechriodus fletcheri* tadpoles breaking free from the foam of the egg mass; (b) an amplexing pair of *Lechriodus fletcheri* adults; (c) *Rheobatrachus silus* from a small stream near Montville, Queensland; (d) *Sphenophryne pluvialis* from a north Queensland rainforest; (e) *Rana papua* in the Mulgrave River, north Queensland; (f) stringy egg mass of *Bufo marinus*. All photographs by courtesy of Dr Gordon Grigg, University of Sydney.

mon in the Cape York area. This species also occurs in Papua New Guinea, New Britain and the Solomon Islands.

Finally in the systematic résumé, reference must be made to the introduced family, the Bufonidae. It is represented by the cane toad (*Bufo marinus*), a species brought into Queensland in 1935 to control certain beetle pests of sugar cane. Its efficacy is this regard remains open to question, but there can be no question of its nuisance value as a pest now widespread throughout eastern Queensland and northeastern New South Wales (Fig. 6.3). Within this region

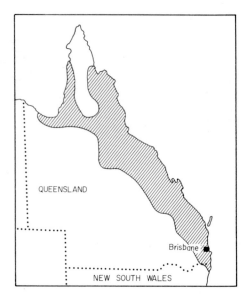

Fig. 6.3 Distribution of the cane toad (*Bufo marinus*) in Australia. After Tyler (1980).

it breeds in ephemeral waters and deposits very high numbers of eggs at each spawning (Fig. 6.2(f)).

Some of the more important ecological adaptations of frogs relate to reproduction; and since most frogs have aquatic larvae such adaptations are most appropriately considered first. An important initial point, however, is that there is so much variation in reproductive patterns between frogs that brief attempts to be comprehensive are likely to fail; all that can be attempted here is a generalization of the principal breeding pattern, and the provision of a few examples of notable deviations from it.

The 'primitive' generalized breeding pattern is that a male begins to produce characteristic mating calls from a body of water, and attracts a female. On approach she is embraced by the male (the pair is then said to be in amplex, see Fig. 6.2(b)), and subsequently deposits her jelly-coated eggs in water (e.g. Fig. 6.2(a)). The eggs are fertilized by the male as they leave the female's body. Eventually the eggs give rise to free-living and voraciously feeding aquatic tadpoles which, in the course of time, develop legs and lose their tails to produce adults. Adults leave the water and emerge on to dry land. Usually the eggs when laid are in a jelly-like mass but frog spawn sometimes assumes the appearance of foam—as in *Limnodynastes* spawn. In *Limnodynastes dorsalis*, for example, the female produces the foam with her front legs; these are used to direct a stream of water beneath her body, and this stream entraps bubbles which on reaching the egg mass and mucous-like jelly behind the female form a mass of foam.

Not surprisingly in a country as arid as Australia, there are many frog species that have compressed the time when free water is required for tadpoles by delaying the actual time of emergence from the egg capsule and shortening larval life. Some, indeed, have suppressed the need for water altogether, although contrary to expectation this has not occurred in most—if not all—desert species; almost all of these lay eggs in water and have aquatic tadpoles. An exception is *Arenophryne rotunda* which inhabits coastal sand dunes in the Shark Bay area of Western Australia; it seems unlikely to have a tadpole stage. Another breeding pattern contrary to expectation is that displayed by *Sphenophryne palmipes* which though not strictly an Australian species—it lives in Papua New Guinea—is closely related to Australian microhylids. This species is unquestionably aquatic as an adult, yet breeds on land. The most bizarre breeding adaptation is undoubtedly that of *Rheobatrachus silus* (Fig. 6.2(c)), the aquatic leptodactylid mentioned above. Eggs are swallowed by the female and develop within her stomach. Neither tadpoles nor female appear to feed during the gestation period. A summary of life-history strategies is given in Table 6.2, which indicates that a variety of

Table 6.2 A summary of life-history strategies adopted by Australian frogs. Table derived from one by Tyler, Watson and Martin (1981).

Eggs	Tadpoles	Family
Foam nest absent		
Aquatic	Aquatic	All (?) Hylidae
		All Ranidae
		Many Leptodactylidae
Terrestrial	Aquatic	Some Leptodactylidae
Terrestrial	Terrestrial	Some Leptodactylidae
		All Microhylidae
Foam nest present		
Aquatic	Aquatic	Some Leptodactylidae
Terrestrial	Aquatic	Some Leptodactylidae
Terrestrial	Terrestrial	Some Leptodactylidae
Parental carriage		
Terrestrial	Male pouch	Some Leptodactylidae (*Assa*)
Terrestrial (?)	Female stomach	Some Leptodactylidae (*Rheobatrachus*)

breeding patterns is displayed by Australian frogs, many of which are interesting adaptive deviations from the primitive pattern. Space precludes further discussion but attention is drawn to the early systematic review of anuran life-histories by Angus Martin (1967). Though more information is now available, this account provides an excellent summary of and introduction to the subject.

Of course all adults have a physiological need for water, whether or not breeding patterns have been altered to obviate its direct need by larvae. Several adaptations have evolved in response to this need, with maximum adaptations being displayed by desert frogs, and least by aquatic or semi-aquatic forms. They may roughly be categorized as structural, behavioural and physiological. Structural responses include the development of digging tubercles for burrowing, modifications to the skin on the upper surface, and the formation of a reasonably impervious outer cocoon. Behavioural responses are many: aggregation of individuals to cut water losses from exposed skin, nocturnal habits, and burrowing are the more obvious. The burrowing desert frogs are particularly noteworthy in this respect (e.g. species of *Cyclorana, Neobatrachus, Limnodynastes*); such frogs can aestivate (lie dormant) at depths of over one metre for two or more years. The best known of them, the water-holding frog (*Cyclorana platycephalus*), digs a subterranean burrow wherein it lies bloated with water contained in subcutaneous lymph sacs and an extended bladder. A prime physiological response has been the evolution of tolerance to significant loss of body water. The moaning frog (*Heleioporus eyrei*) from southwestern Western Australia, for example, experiences an average water loss of about one quarter of its body weight while foraging at night in summer (a deficit made good not by drinking but from water in its food and from soil water). Another species, *Litoria caerulea*, can survive after losing almost half of its body weight in water.

In general, adult frogs have catholic tastes in food, and they have been aptly described as 'unspecialized opportunistic predators'. Insects, nevertheless, appear to predominate as a food item. Additional food items noted include other frogs, fish, and a variety of invertebrates apart from insects. Little precise information is available concerning larval feeding habits and food in Australian species, a rather surprising fact in view of the obvious importance of food-gathering in larval ecology (and therefore in subsequent adult population dynamics), and the fact that tadpole anatomy is so profoundly modified for the collection and processing of food. In general, vegetable detritus, fine particulate organic matter, plant material and other tadpoles seem important food materials for most tadpoles though some ingest a greater proportion of animal food; for example, tadpoles of Fletcher's frog (*Lechriodus fletcheri*; Fig. 6.2(a)) are said to be habitual cannibals and carnivores and this is probably an adaptation to life in temporary pools, a habitat typical for the species. The same may also be true for *Pseudophryne occidentalis* which also occurs in temporary pools quite devoid of large aquatic plants. Tadpole mouths are frequently elaborately structured, and sometimes used for purposes other than feeding. In the stream-dwelling tadpoles of Lesueur's frog (*Litoria lesueurii*) the mouth has numerous labial papillae forming a suction cap and is used as an adhesive structure.

The major predators on adult frogs appear to be snakes, and then birds. Various mammals and even some spiders also feed on frogs. Tadpoles fall prey to fish, various water bugs (Hemiptera), beetle adults and larvae, aquatic spiders (*Dolomedes*), tortoises, snakes, birds, certain mammals, and other tadpoles.

Finally, it is appropriate to remark briefly on the impact of man on the Australian frog fauna. Both beneficial and deleterious impacts are discernible. Amongst beneficial impacts is included the local provision of new water-bodies in many arid and semi-arid regions where standing water would either not occur or occur only sporadically. Such waters have undoubtedly been used to advantage by several species. Unfortunately, the deleterious impact of man is likely to have had greater significance. Principal events in this category include drainage of wetlands, deforestation, and the introduction of exotic animals. *Gambusia affinis, Cyprinus carpio* and *Bufo marinus* are amongst the most important introductions in this regard, but the continued annual importation of many millions of individuals of aquarium fish (see chapter 5 for further details) is a threat to Australian frogs that can scarcely be underestimated. The importation by aquarists of European newts poses a similar threat (newts are urodele amphibians with aquatic larvae and terrestrial adults). It is not surprising that in the face of so many threats several Australian frog species are considered as endangered (Fig. 6.4). The full extent of man's deleterious impact will never of course be determined.

One further human impact may be noted: an extension of natural ranges following accidental transport by man. The isolated occurrence in northwestern Western Australia of *Limnodynastes tasmaniensis* some 2000 km from its normal area of distribution (eastern and southeastern Australia) provides a clear example. It seems that this frog was introduced to the northwest from Adelaide by carriage in the foundations of transportable housing constructed in South Australia.

REPTILES

Tortoises (turtles), crocodiles, and some lizards and snakes constitute the reptile fauna of Australian fresh waters. A recent and comprehensive account of these groups has been given by Cogger (1975).

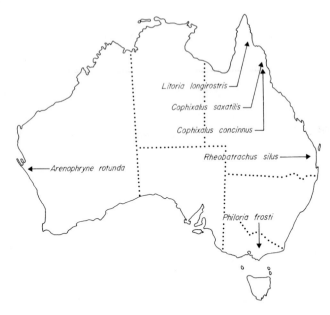

Litoria longirostris —

Cophixalus saxatilis —

Cophixalus concinnus —

Rheobatrachus silus —

— Arenophryne rotunda

Philoria frosti

Fig. 6.4 Distribution of endangered frog species in Australia. After Tyler (1979).

There are three families of lizards, the Agamidae (dragons), Varanidae (monitors) and Scincidae (skinks). The species involved are listed in Table 6.3. All 10 species listed are found associated with vegetation and rocks in or beside streams, creeks, swamps and rivers, and, in a few cases, mangrove swamps. Several other species not listed are less closely associated with inland waters. Those in Table 6.3 range in habit from what can best be regarded as semi-aquatic to those which must be regarded as full members of the aquatic fauna. Most of these when disturbed and in an appropriate location will drop into water and remain immersed. Fish, frogs, and insects form the major part of their diet. Most aquatic lizards are northern forms (Fig. 6.5), but at least three, the eastern water dragon (*Physignathus lesueurii*), the eastern water skink (*Sphenomorphus quoyii*), and the southern water skink (*S. tympanum*), occur widely in southeastern Australia.

The eastern water dragon (Fig. 6.6) has two subspecies, a larger northern one which can reach a length of 90 cm and a smaller southern one, 60 cm. The southern form, *P. lesueurii howittii*, is sometimes referred to as the Gippsland crocodile or Gippsland water lizard. Its subspecific name was proposed by Professor McCoy (1884) whose coloured plate of the animal forms the basis of Fig. 6.6. The eastern water skink is a much smaller lizard (up to 10 cm long)

Table 6.3 Aquatic lizards and snakes.

Family	Scientific name	Common name
Lizards		
Agamidae	*Lophognathus temporalis*	–
	Physignathus lesueurii	Eastern water dragon
Varanidae	*Varanus indicus*	Mangrove monitor
	V. mertensi	Merten's water monitor
	V. mitchelli	Mitchell's water monitor
	V. salvator	Asian water monitor
	V. semiremex	Rusty monitor
Scincidae	*Sphenomorphus kosciuskoi*	Alpine water skink
	S. quoyii	Eastern water skink
	S. tympanum	Southern water skink
Snakes		
Boidae	*Liasis fuscus*	Water python
Acrochor-didae	*Acrochordus javanicus*	Javan file snake
Colubridae	*Amphiesma mairii*	Keelback or water snake
	Enhydris punctata	Spotted water snake
	E. polylepis	Mcleay's water snake
	Stegonotus cucullatus	Slaty-grey snake
Elapidae	*Austrelaps superbus*	Copperhead
	Hemiaspis signata	Black-bellied swamp snake
	Pseudechis porphyriacus	Red-bellied black snake

usually to be found basking on rocks and logs beside creeks and rivers, though sometimes also at a considerable distance from water. It has a catholic taste in food and produces live young, two to three per litter. The southern water skink is another small lizard (up to about 8 cm long) and like the eastern water skink may also be found a considerable distance from water; it occurs mainly along watercourses or in forest clearings near water. It too produces live young. It is believed that two species are presently confused under the single name of *S. tympanum*.

Before leaving the lizards, brief mention may also be made of one species not listed in Table 6.3: the Lake Eyre dragon (*Amphibolurus maculosus*). It is mentioned because it lives in what must surely be the most inhospitable of all 'aquatic' habitats – the barren salt-encrusted edges of dry saline lakes in the Lake Eyre basin.

Several snakes occur closely associated with fresh waters in Australia, besides several more which are less closely associated. The aquatic and semi-aquatic species are listed in Table 6.3. The latter group, in particular, has been rather arbitrarily derived. The distribution of the species listed is indicated in Fig. 6.7.

D

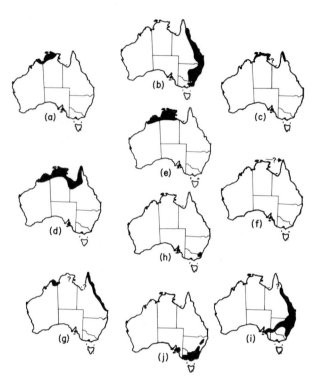

Fig. 6.5 Distribution of aquatic lizards. (a) *Lophognathus temporalis*; (b) *Physignathus lesueurii*; (c) *Varanus indicus*; (d) *V. mertensi*; (e) *V. mitchelli*; (f) *V. salvator*; (g) *Varanus semiremex*; (h) *Sphenomorphus kosciuskoi*; (i) *S. quoyii*; (j) *S. tympanum*. Compiled and redrawn from Cogger (1975).

The elapids are quite dangerous, but are usually shy animals, and a good pair of rubber boots and unstealthy habits near water are probably sufficient protection for most freshwater biologists. Like the aquatic lizards, several aquatic snake species occur only in northern Australia—a predictable event since aquatic snakes need a warm milieu for they cannot bask in the sunshine to raise body temperatures as can terrestrial snakes. However, four semi-aquatic species may be encountered in the east and southeast. The keelback or freshwater snake (*Amphiesma mairii*) is widespread in both northern Australia and the eastern part of Queensland; it feeds mainly on frogs. The three snakes most likely to be found in eastern and southeastern swamps are the cop-perhead, black-bellied swamp snake, and the red-bellied black snake. All are venomous.

The fully aquatic snakes show several interesting adaptations to the

Fig. 6.6 The eastern water dragon. Reproduced from the original colour plate in the *Prodromus of the Zoology of Victoria* (McCoy 1884).

freshwater environment. These have been studied best in *Acrochordus javanicus*, a non-venomous species. In this snake the adaptations include dorsal nostrils, a mechanism to seal the nostrils when submerged, a well-developed and richly vascularized lung, long non-ventilatory (non-breathing) periods, an effective elimination of carbon dioxide through the skin, and a low metabolic rate. Further, since reptilian eggs cannot develop in water, the species bears live young.

There are two sorts of crocodile in Australia, the large estuarine crocodile (*Crocodylus porosus*, Fig. 6.8) and the smaller freshwater crocodile or Johnston's crocodile (*C. johnstoni*, Fig. 6.9). Both are confined to a broad northern arc stretching from the northwest of Western Australia to the northeast coast of Queensland. The estuarine crocodile, despite its common name, is frequently found considerable distances upstream from estuaries—there are records of some found 100 km from river mouths—as well as far out to sea. This species occurs widely throughout the Indonesian archipelago and southeast Asia. The freshwater crocodile, on the other hand, is endemic to

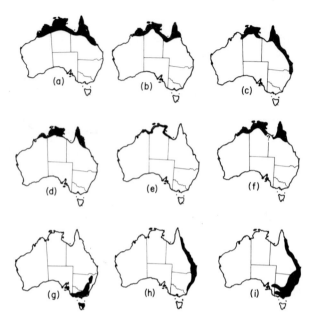

Fig. 6.7 Distribution of aquatic snakes. (a) *Liasis fuscus*; (b) *Acrochordus javanicus*; (c) *Amphiesma mairii*; (d) *Enhydris polylepis*; (e) *E. punctata*; (f) *Stegonotus cucullatus*; (g) *Austrelaps superbus*; (h) *Hemiaspis signata*; (i) *Pseudechis porphyriacus*. Compiled and redrawn from Cogger (1975).

Fig. 6.8 Head of juvenile *Crocodylus porosus*, the estuarine crocodile. Photograph by courtesy of Dr Gordon Grigg.

Fig. 6.9 *Crocodylus johnstoni*, Johnston's crocodile. A captive specimen photographed at James Cook University, Zoology Department.

Australia, and is confined to fresh water. Formerly, both species were common in Australia, but following extensive hunting for commercial purposes after the second world war many populations became critically depleted. From a conservation viewpoint the two species are now regarded as in a vulnerable position and are protected by legislation throughout their ranges. The present status for *C. porosus* is that (a) in the Northern Territory, populations are recovering slowly in some tidal rivers but elsewhere are only steady or falling, (b) Western Australian populations appear to be recovering slowly, and (c) in Queensland, populations are generally still falling and are dangerously low in some areas. For *C. johnstoni*, conservation prospects are perhaps not quite so gloomy; although the status of the species remains uncertain, it appears to be moderately common in many parts of its range.

The estuarine crocodile is said to attain lengths of about 7 m, and though few appear to survive to such a length, this species is quite large enough to be dangerous in certain situations. Fortunately for its own survival, it is a shy and wary beast. Its normal food consists of fish, together with various reptiles, waterbirds and such mammals as approach too near the river bank. The wet season (November–March) is the breeding season when some 50 or so eggs are laid in a shallow hole high on river banks. The nests are covered by soil and

leaves to form a mound and the eggs hatch after about three months' incubation. Flooding kills the embryos. Apparently, adult females remain near the
nest during incubation and with the hatchling group for up to 2½ months.
Considerable variation in population densities occurs both within a single river
and between rivers. This has been attributed to adult movement and
differences in breeding success. And over a wide salinity range the species can
regulate within narrow limits both the concentration and composition of its
body fluids.

Most of the above information is of recent origin and derives from an intensive investigation of *C. porosus* by Professor Harry Messel and Drs Webb
and Grigg and associates at the University of Sydney. Much less, unfortunately, is known about the ecology and physiology of Johnston's crocodile. This
species is smaller and rarely attains a maximum length of 3 m. It does not
attack man, and despite its forbidding teeth row is said to be quite innocuous.
A nocturnal animal which heats faster than it cools, it feeds on fish, frogs,
crustaceans, lizards and small mammals and birds. It breeds in spring (October–November), that is, near the end of the dry season. About 20 eggs are
laid in holes on the river bank.

Both Australian crocodiles display several interesting adaptive features
which may briefly be noted. The hind feet are webbed, but provide motive
power for only slow movement; it is the powerful tail which provides the
means for rapid movement. The nostrils have valves that seal them when
submerged and there is a secondary palate enabling breathing to occur with the
mouth submerged, an adaptation aided by another valve at the end of the
secondary palate and the base of the tongue. The ventricular chamber of the
heart is completely divided enabling separation of venous and arterial blood
(though some mixing occurs), and the lungs are highly developed. These and
other adaptations are clearly of high value indeed for crocodiles have survived
as a group since the Triassic period, approximately 180 million years ago.

The freshwater tortoises (or turtles) belong to an even more ancient
group: the order Chelonia containing them is known from Permian times,
some 210 million years ago. Those in Australia belong in two families, the
Chelidae and the Carettochelyidae. The Chelidae is the dominant family and is
characterized by being pleurodirous (the head and neck are bent beneath the
front of the shell by one or more horizontal folds), hence the name 'side-
necked tortoises'. It is entirely freshwater in habit and confined to Australia,
Papua New Guinea and South America. The Carettochelyidae is cryptodirous
(the head and neck are bent beneath the shell by vertical folds) and known only from northern Australia and Papua New Guinea.

Much awaits discovery in this interesting group of reptiles, not least with
regard to its taxonomy. At present four genera of Chelidae are recognized with

Fig. 6.10 *Chelodina longicollis*, long-necked tortoise or eastern snake-necked turtle. Photograph by courtesy of Dr B. C. Chessman.

a total of some 15 species. The genus *Chelodina* (long-necked tortoises) has six species of which only two, however, are found in southeast Australia, namely the broad-shelled river tortoise (*C. expansa*) and the eastern snake-necked or long-necked tortoise (*C. longicollis*, Fig. 6.10). The remaining three genera are all short-necked. *Elseya* has six (or, perhaps, five) species, of which one, the Murray tortoise (*E. macquarii*, Fig. 6.11), can be regarded as common in the southeast. And *Pseudemydura* has a single species (*P. umbrina*) known only from near Perth where it is regarded as in imminent danger of extinction despite complete protection.

For present purposes tortoise ecology can be generalized in the statement that typically tortoises are inhabitants of slow-moving rivers, billabongs and associated swamps, their diet comprises fish and crustaceans, molluscs and other suitably sized invertebrates, and most appear to be summer breeders

Fig. 6.11 *Emydura macquarii*, Murray tortoise. Photograph by courtesy of Dr
B. C. Chessman.

with eggs (up to 20) in chambers high on river banks. Some species, at least,
are capable of extensive overland migration.

    The other Australian freshwater tortoise family contains a single species,
the pitted-shell turtle (*Carettochelys insculpta*), confined to Papua New
Guinea and a small area of the Northern Territory. It was first recorded from
Australia in 1969, and is found in both the estuarine and freshwater reaches of
northern rivers, in large waterholes associated with these, and in billabongs
isolated during the dry season. It feeds upon molluscs, plant material and fish.

7 Waterbirds

The only birds usually regarded by ornithologists as 'waterfowl' are those in the order Anseriformes: ducks, swans and geese. There is a similar restriction with the word 'waders': for ornithologists, waders are members of the order Charadriiformes — snipe, plovers, dotterels, gulls and allies. But there are many birds associated with water that belong to neither the Anseriformes nor the Charadriiformes. To cover all the avifauna associated with inland waters, the term 'waterbirds' is used here.

As a group, Australian waterbirds have not been dealt with at all comprehensively or thoroughly, and the present chapter cannot hope to provide any remedy. All that is attempted is a brief introduction to the various orders of birds associated with Australian inland waters, a discussion of some salient points concerning their significant members, and a general comment on waterbird habitats.

BIRD ORDERS ASSOCIATED WITH AUSTRALIAN INLAND WATERS

Nine bird orders are represented in the Australian inland aquatic avifauna. Their composition and diversity is indicated in Table 7.1. This table is no more than a reasonable approximation, for in compiling it there were several difficulties: when does a species' habitat become sufficiently 'aquatic' to merit inclusion of the species; or when should a vagrant species be included. There is, furthermore, a certain lack of taxonomic agreement amongst ornithologists. The table will suffice for present purposes.

Perhaps the Passeriformes should also have been included, for, although the fully aquatic dipper (*Cinclus cinclus*) of European streams is absent from Australia, there are a few Australian passerines that do live close to inland waters: the widespread reed warbler (*Acrocephalus stentoreus*), yellow chat (*Ephthianura crocea*) and little grassbird (*Megalurus gramineus*), all of which inhabit reed beds, and the purple-crowned wren (*Malurus coronatus*), are notable examples. Also not in the table are the loons (order Gaviiformes) and

Table 7.1 Composition of Australian inland aquatic avifauna.

Order	Scientific name	Family Common name	No. genera	No. species
Podicipediformes	Podicipedidae	Grebes	3	3
Pelecaniformes	Pelecanidae	Pelicans	1	1
	Anhingidae	Darters	1	1
	Phalacrocoracidae	Cormorants	1	4
Ciconiiformes	Ardeidae	Herons, egrets, bitterns	5	13
	Ciconiidae	Storks	1	1
	Threskiornithidae	Ibises, spoonbills	3	5
Anseriformes	Anatidae	Ducks, geese, swans	13	19
Falconiformes	Accipitridae	Eagles, harriers	2	2
	Pandionidae	Ospreys	1	1
Galliformes	Phasianidae	Quails	1	1
Gruiformes	Gruidae	Cranes	1	2
	Rallidae	Rails, crakes, hens, coot	7	13
Charadriiformes	Jacanidae	Jacanas	1	1
	Rostratulidae	Painted snipe	1	1
	Charadriidae	Plovers, dotterels	3	7
	Scolopacidae	Curlews, snipe, sandpipers, stints	8	17
	Recurvirostridae	Avocets, stilts	3	3
	Laridae	Gulls, terns	5	9
Coraciiformes	Alcedinidae	Kingfishers	2	3
		Totals	64	107

the flamingoes (family Phoenicopteridae). They are absent because neither group occurs in Australia. The absence of the loons is not surprising for they are basically an Holarctic group, but the absence of flamingoes is perhaps unexpected. Flamingoes did occur in Australia earlier, and fossils as recent as the Pleistocene have been collected. It has been suggested that they became extinct because the increased aridity could not provide the required degree of permanency in the large water-bodies that flamingoes usually frequent.

Grebes (order Podicipediformes)

The grebes are large to medium-sized waterbirds with lobed and partly-webbed feet. They rarely fly during daylight. There are three Australian species in three genera, viz *Podiceps, Poliocephalus* and *Tachybaptus*; one of them, the hoary-headed grebe (*Poliocephalus poliocephalus*, Fig. 7.1(a)), is endemic. Grebes are most frequently found on inland water-bodies, but are occasionally found in estuaries and other sheltered marine waters. Food is collected by diving and foraging on the bottom.

Pelicans, darters, cormorants (order Pelecaniformes)

The Australian pelican (*Pelecanus conspicillatus*, Fig. 7.1(b)) is a large, striking waterbird with a white body and black wings and tail. It is characterized by its long beak from which hangs a large pink pouch in which up to 14 litres of water can be held. The pouch probably serves a thermoregulatory function. The pelican is found throughout Australia, but nesting colonies are chiefly southern. They are located either in swamps, on small islands, or on sandbars, and the nest is usually a slight hollow (a scrape) in the ground surrounded by plant material added later. Occasionally, nests are in bushes. Large colonies nest near the Coorong and Lake Alexandrina, South Australia, but wherever the species breeds, an undisturbed nesting area for at least 3 months is required. Pelicans, like many waterbirds discussed below, are found in both coastal marine and inland aquatic situations. Surprisingly for such large, ungainly birds, they perch in trees.

Related to the pelican are the much smaller darter (*Anhinga melanogaster*) and cormorants (*Phalacrocorax* spp., Fig. 7.1(c)). Both are diving birds with long necks and bills, short legs, and webbed feet. The cormorants are more common and are distinguished by the terminal hook on their beaks. Four Australian cormorants are found inland (as well as coastally); another is a strictly marine species. In general, fishermen do not like cor-

Fig. 7.1 (a) Hoary-headed grebe; (b) pelican; (c) big black cormorant; (d) white egret; (e) straw-necked ibis; (f) wood duck. All photographs by courtesy of Mr T. Lowe.

morants, but recent research has shown that cormorant depredations on fish are not as significant as previously assumed. In lakes of inland New South Wales, for example, the little black cormorant (*P. sulcirostris*) feeds mainly on exotic fish (carp, redfin, mosquito fish), and the pied cormorant (*P. melanoleucos*) feeds mainly on native freshwater crustaceans.

Herons, ibises, spoonbills, etc (order Ciconiformes)

This group of waterbirds is characterized by having very long legs, feet that are not webbed, a long neck, and a large beak. It includes a wide variety of marsh and swamp birds. Basically, there are three sorts: the herons, egrets (Fig. 7.1(d)) and bitterns; the black-necked stork or jabiru; and the ibises and spoonbills. The first two sorts are distinguished by their long, pointed and straight beaks; the ibises have long and pointed beaks that are downcurved, and the spoonbills long, straight but distally flattened beaks. As a general rule, members of this group feed in shallow lake margins, swamps, marshes and river pools, are mostly carnivorous, and nest colonially in trees near water. Several, however, obtain food elsewhere and breed in rushes. In this connection, it is noted that the straw-necked ibis (*Threskiornis spinicollis*, Fig. 7.1(e)) is useful in pest-control; it feeds on grasshoppers, crickets and caterpillars from pasture, though the overall effect on locusts is apparently not significant. Ibises in southern Australia tend to breed regularly in spring, but, when favourable (flood) conditions occur in central Australia, breeding there may take place at any time. The nomadism of ibises is an important element in this use of less predictable breeding sites.

Ducks, swans (waterfowl) (order Anseriformes)

Of all waterbirds, waterfowl have received most attention due largely to their popularity with sportsmen. However, considering the area of Australia, the diversity of its waterfowl is very low: there are only 19 native species (cf. the British Isles with 47). Nevertheless, what Australian waterfowl lack in diversity is compensated for in biological interest: several endemic species and six monotypic endemic genera occur. The latter are: *Cereopsis novaehollandiae*, the Cape Barren goose; *Anseranas semipalmata*, the magpie goose; *Stictonetta naevosa*, the freckled duck; *Malacorhynchus membranaceus*, the pink-eared duck; *Chenonetta jubata*, the wood duck (Fig. 7.1(f)); and *Biziura lobata*, the musk duck (Fig. 7.2(a)). Perhaps the most interesting of these is the magpie goose. This differs in so many structural features from all other Anseriformes that it is placed in its own subfamily, the Anseranatinae. All remaining Australian waterfowl are placed either in the subfamily Anserinae (swan (Fig. 7.2(b)), Cape Barren goose, freckled duck, whistle-ducks) or Anatinae (all other ducks). Incidentally, there are no true geese in Australia, despite the vernacular use of the term for some Australian ducks.

Although many Australian waterfowl are nomadic, and many occur widely, it is possible to classify them on the basis of three generalized distributional patterns: a northern pattern of distribution (magpie goose, the two species of whistle-duck, the Burdekin duck, and the two species of pigmy geese); a southern pattern (Cape Barren goose, black swan, mountain duck, chestnut teal, freckled duck, shoveler, musk duck, and blue-billed duck); and a continental pattern (grey teal, pink-eared duck, black duck, white-eyed duck and wood duck).

Fig. 7.2 (a) Musk duck; (b) black swan; (c) black-tailed native hen; (d) coot; (e) red-capped dotterel; (f) black-fronted dotterel. All photographs by courtesy of Mr T. Lowe.

A slightly different way of viewing waterfowl distribution is to consider the distribution of waterfowl habitats. On this basis, Australia can be divided into four waterfowl regions, each with characteristic species.

There is a *Central Region* where nomadism prevails. The grey teal (*Anas gibberifrons*), pink-eared duck (*Malacorhynchus membranaceus*), and wood duck (*Chenonetta jubata*) are typical ducks of this region.

There is a *Murray–Darling Region*, the most prolific of all waterfowl breeding areas, with four distinct breeding subregions (Riverina, Macquarie–Gwydir, Bulloo–Paroo, and Lower Darling). It is here that duck ecology has been most intensively studied. Species regarded as typical for the region are the black duck, pink-eared duck, grey teal, white-eyed duck, wood duck and black swan. Musk duck and blue-billed duck are often numerous also.

The third region is the *Southern Region*. Five species characterize it: the mountain duck, chestnut teal, wood duck, Cape Barren goose and black swan. The Cape Barren goose is confined to it, and the black swan, though not confined, is certainly more common in it than elsewhere.

And the fourth region is the *Northern Region*. Six waterfowl species are virtually restricted to it, and another nine breed there. The green and white pigmy geese (*Nettapus* spp.) are the most aquatic of all Australian waterfowl and rarely leave the northern billabongs and lagoons to which they are confined. The magpie goose is now similarly confined, but formerly lived as far south as Victoria. It has been extensively studied because of its agricultural impact: flocks eat newly sown seed and are otherwise a nuisance. The two species of whistle-duck (*Dendrocygna* spp.) and the Burdekin duck, likewise characteristic of the region, have also been studied.

The above account of waterfowl distribution in Australia may have left the impression that the species involved do not move much within the continent. If that is true, then correction is needed: most Australian waterfowl do undertake considerable journeys. However, they do not migrate in the sense that many northern hemisphere wildfowl do, that is, undergo regular, precisely-timed, seasonal movements in a predictable direction. Australian waterfowl 'migration', such as it is, is much less predictable in both timing and direction: Australian waterfowl are better described as nomads than migrants. The usual pattern seems to be that most disperse from their breeding or feeding grounds whenever these become unsuitable, and may never return. The regular annual phenomena of changing daylength and temperature, so important in controlling northern hemisphere waterfowl migration, are less important in Australia where rainfall and the availability of food are more important factors. Undoubtedly, this feature is an adaptive behavioural response to climatic aridity and variability. Such directional biases as do occur in waterfowl movement relate to a southeastwards drift in summer, and, in northern species, an east–west (winter–summer) movement.

Similarly, the breeding of Australian waterfowl is less tied to precisely-timed environmental phenomena than it is for waterfowl in the northern hemisphere. For many northern hemisphere species, photoperiodism plays an important role in initiating breeding. But long days without water are of no use to waterfowl, and a major controlling factor for many Australian species is one with more predictive usefulness for the occurrence of suitable breeding conditions: rainfall. Thus, in southern Australia (including the southern part of the Murray–Darling basin), where rain falls mainly in winter, waterfowl breed mostly in late winter and spring; in the north, where summer rains are the rule, they breed mostly in late summer and autumn; and in central Australia, where rain falls erratically, breeding is erratic in timing and follows rainfall or flood whatever the season. The major stimulus appears to be not the rain itself, but rather the effect of rain on water levels. In any event, for many species, there is a close correlation between habitat water level and the onset of breeding. In at least some species, however, an endogenous cycle and a photoperiodic response provide a firm predictive element in the timing of breeding (e.g. for the black duck). Setting aside the actual mechanisms that time the onset of breeding, two broad categories may nevertheless be recognized in so far as breeding season predictability is concerned. In one category — which includes most species — reproductive cycles display considerable regularity. In the other category, the timing of the reproductive cycle shows facultative adaptation to local conditions; species characteristic of the Central Region fall into this category.

Actual breeding sites vary in position. According to location, five sorts of nest may be distinguished: those in swamp vegetation; open nests on land; open nests on raised sites such as tree stumps; nests concealed under rocks and other material on the ground or concealed in holes in the ground; and holes in trees. There is a good deal of overlap in these locations as to species using them, though some species have quite specific preferences. The black duck is one that has not; its nests are found unconcealed on land, on raised tree stumps, and in tree-holes. The magpie goose, on the other hand, nests only in swamp vegetation. Irrespective of nesting site, most nests are simple hollows; species that build a more elaborate nest are the magpie goose, swan (sometimes), white-eyed duck, freckled duck, musk duck, and blue-billed duck. Many different sorts of inland aquatic habitat are used for breeding purposes, but six major sorts have been distinguished: (a) seasonal swamps, (b) coastal lakes, estuaries and dams, (c) permanent swamps and lakes, (d) semi-permanent swamps and lakes, (e) inland watercourses and dams, and (f) flood waters. Breeding regularity is greatest in habitats (a)–(c), and least in (d)–(f).

The food of most Australian waterfowl is predominantly plant material, though several species include invertebrates in their diet and four species are predominantly carnivorous (Burdekin duck, shoveler, pink-eared duck, and musk duck). A very general idea of the importance of various food items to inland species (mainly those of the Murray–Darling area) and tropical species is indicated in Fig. 7.3. As can be seen, sedges and grasses are of major importance for both groups, but the inland species consume relatively more terrestrial plant material and greater numbers of animals than do the tropical species. Most food is collected in or at the edge of the water-body, but some species make extensive use — sometimes on a seasonal basis — of terrestrial vegetation. In the north, these are the magpie goose, the grass whistle-duck, and the wood duck; in the south, they are the swan, Cape Barren goose, and

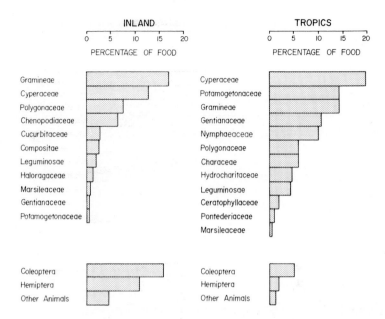

Fig. 7.3 Average diet of waterfowl in tropical coastal areas, and in southern inland areas (mainly Murray-Darling basin). Top scale indicates total percentage of food. Redrawn from Frith (1977).

mountain duck. These species graze pasture and crops and, depending on the extent to which they do this, may cause significant damage. Species more closely tied to water for feeding clearly occupy a number of trophic niches: some dive, others do not, some feed only at the edge, others only at the surface. Because of this, it seems that in mixed species flocks feeding at one location, competition does not occur until times of drought when food is scarce and when individual preferences are less important.

Major impacts on wildfowl populations in Australia include habitat destruction, hybridization with introduced species (for some), persistence of certain pesticide residues in the body, and over-exploitation by hunting. The last has had particular impact in recent years on some of the less common but biologically most interesting species. Despite an intensive hunter education programme by the local National Parks and Wildlife Service before the opening of the 1980 duck season, for example, it is estimated that hunters shot between 500 and 1000 specimens of the freckled duck—an endangered and totally protected species—at Bool Lagoon, South Australia. It is clear that effective management of wildfowl resources has not yet been achieved in Australia.

Eagles, harriers, ospreys (order Falconiformes)

Three birds of prey may be considered as waterbirds in that they habitually or frequently hunt near inland waters. Two of them, the white-breasted sea-eagle and the osprey, are not confined to inland waters but occur also at the coast and near estuaries. One of them, the swamp harrier, is more inland in distribution. It does, however, range considerable distances away from water to drier areas. All are distributed widely in Australia.

Quails (order Galliformes)

Most members of the Galliformes are quite terrestrial in habitat. However, the king quail (*Coturnix chinensis*) inhabits swampy areas, marshy ground and wet grassland, and on this account deserves passing mention. It is distributed from the Kimberleys along a broad coastal strip as far south as Victoria.

Cranes, rails, crakes, hens, coots (order Gruiformes)

The cranes are large marsh birds with long necks and legs. There are two in Australia, the brolga (*Grus rubicunda*) recorded widely in the north and eastern half of the continent, and the sarus crane (*G. antigone*) of restricted northern distribution. The brolga is best known for its elaborate courtship dance. Though mainly vegetarian and mostly feeding on *Eleocharis* bulbs from swamp mud, brolgas do eat various animals and graze on pasture. Pairs nest alone.

The remaining waterbirds of this group are within the rail family, the Rallidae. Small to medium-sized, generally hen-like birds, they usually occur in reed beds. Several swim strongly, and a few dive for food. The most common are the marsh (Baillon's) crake (*Porzana pusilla*), spotted crake (*P. fluminea*), spotless crake (*P. tabuensis*), black-tailed native hen (*Gallinula ventralis*, Fig. 7.2(c)) and its Tasmanian cousin (*G. mortierii*), the moorhen (*G. tenebrosa*), the swamp hen (*Porphyrio porphyrio*), and the coot (*Fulica atra*, Fig. 7.2(d)). Several, though shy, are common in ornamental urban lakes and are well-known; it is surprising that we are so ignorant about the ecology of some of them. However, a good deal is known about the Tasmanian native hen because it sometimes damages crops. Unlike the mainland form, the Tasmanian native hen cannot fly, a reflection of the different type of water-body inhabited by these two species; whereas the Tasmanian native hen lives near permanent water-bodies, the nomadic mainland one typically inhabits temporary waters. The ecological equivalent of the Tasmanian form in mainland permanent waters appears to be the moorhen.

Jacanas, dotterels, stilts, gulls, terns, etc (waders) (order Charadriiformes)

More species of Australian waterbirds belong in this group, the waders, than any other; about 40 in some 21 genera are recorded from Australian inland waters (additional ones frequent coastal waters only). Most are found not only near inland waters, but in shallow coastal waters, estuaries, and on wet pasture. Shallow coastal waters seem a preferred habitat for most, though there are a few found only in freshwater habitats (e.g. the jacana).

Six wader families are of interest, with the Scolopacidae (snipe, sandpipers, stints, godwits, greenshanks) the most diverse. Many species of this family, however, rarely occur in Australia, for the family as a whole is notably migratory and all Australian members breed elsewhere and visit Australia only during spring and summer; they arrive in August and depart between March and May (odd individuals and flocks may occasionally linger through a winter). Breeding takes place in the northern hemisphere; the red-necked stint (*Calidris ruficollis*), for example, one species that is common in Australia, breeds in Siberia and Alaska, whilst the curlew sandpiper (*C. ferruginea*), another common form, breeds in the Arctic. Some other wader families include forms that occur here only as migrants so that the total species number of migrant waders in Australia exceeds the number of 'residents' (only about 16 species of waders frequenting Australian inland waters breed here, whereas there are about 24 migrant forms).

With so diverse a group of birds, further discussion must be confined to those species that are of greater interest or abundance than others. The jacana or lotus bird (*Jacana gallinacea*) is certainly one of them. Medium-sized, this waterbird has exceedingly long legs, toes and toenails and can walk on floating plants and other matter at the water surface. Even nests are constructed on floating vegetation, but should these through some mishap become uninhabitable, then eggs and chicks can be moved by parent birds using their wings. Only the jacana is known to do this. It is common on ponds along the eastern coast of Queensland but occurs in northern Australia also, and as far south as Sydney.

Two dotterels deserve mention. The red-capped dotterel (*Charadrius ruficapillus*, Fig. 7.2(e)) is common both at the coast and near salt lakes. At the latter it is said to eat brine shrimp, but more likely eats snails. The blackfronted dotterel (*C. melanops*, Fig. 7.2(f)) is much more an inland species and

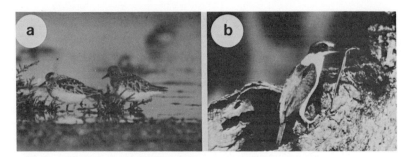

Fig. 7.4   (a) Red-necked stints; (b) sacred kingfisher. Photographs by courtesy of Mr T. Lowe.

frequents margins of shallow pools and creeks. When still, it is often almost perfectly camouflaged.

One of the commonest migrant waders is the red-necked stint (*Calidris ruficollis*, Fig. 7.4(a)). Breeding in Siberia and Alaska, it travels thousands of kilometres yearly, yet is scarcely bigger than a sparrow. It is a bird of coastal and inland aquatic environments.

Fig. 7.5   Silver gull. Photograph by courtesy of Mr T. Lowe.

The two stilts and the avocet of the family Recurvirostridae also travel considerable distances, but always within Australia, as they breed here. However, they do so near shallow, ephemeral salt lakes of central Australia and this means that they need to be nomadic. At least one stilt as well as the avocet is known to respond to unseasonal rain by breeding in much the same way as many nomadic waterfowl.

Finally, no account of Australian waders would be complete without mention of the ubiquitous silver gull (*Larus novaehollandiae*, Fig. 7.5). This extremely common species, an opportunistic and omnivorous feeder, is favoured by the presence of man and huge numbers occur near rubbish tips, sewage works and beach resorts. Research on its movements using bird-banding techniques has shown that young gulls are most mobile but that mobility varies from individual to individual. However, though long distances are sometimes travelled, these do not compare with those travelled by stints, sandpipers and allies nor the stilts, so that these gulls cannot be classed as true migrants nor as nomads; they have been regarded as partial migrants. The species occurs well inland and in all coastal districts. Nesting requirements are for protected, preferably insular breeding sites, where colonies may form. This requirement is a serious constraint for such localities are not common near Australian towns. As a result, during breeding (spring and summer in the southeast), parent birds may need to commute between suitable nesting sites and feeding grounds.

Kingfishers (order Coraciiformes)

Three kingfishers occur near inland waters, the azure kingfisher (*Alcyone azurea*), the little kingfisher (*A. pusilla*), and the sacred kingfisher (*Halcyon sancta*, Fig. 7.4(b)). Several relatives are found in dry forest areas—despite their vernacular name—and in mangroves. All three inland kingfishers also frequent mangroves, and all are carnivorous.

MAJOR TYPES OF WATERBIRD HABITAT

A useful conclusion to this section is to consider waterbirds not in terms of systematic grouping but in relation to major habitat types. Table 7.2 has been compiled to provide a brief summary. In it, six major natural waterbird habitats have been arbitrarily delimited, namely, and roughly in order of increasing 'wetness': floodplain forests, irrigated pastures, meadow swamps and natural wet pastures, reedswamps, open waters, and salt lakes. Small streams

Table 7.2 Some major natural habitat types and their principal waterbirds.

Floodplain forest	Irrigated pasture	Meadow swamps natural wet pasture	Reedswamp	Open water	Salt lakes
Dominant vegetation					
Trees, grass, sedges, rushes	Grass, semi-aquatic macrophytes	Sedges, rushes	Rushes, sedges, reeds	Submerged hydrophytes	None
Feeding area for:					
Ibis, snipe, dotterel, kingfishers	Mountain duck, ibis, herons	Spoonbills, egrets, herons, waterhens, brolga, stilts, ducks, swans, yellow chat	Spoonbills, egrets, swans, coots, ducks, reed warblers, bitterns, crakes, rails, little grassbird, yellow chat	Swans, musk ducks, cormorants, pelicans, darters, hawks, waterhens	Banded stilts, avocets, red-capped dotterels, red-necked stints, curlew, sandpipers
Breeding area for:					
Ducks, cormorants kingfishers	Plovers, dotterels	Ducks, crakes, rails, little grassbird, yellow chat	Cormorants, ibises, swans, grebes, bitterns, reed warblers, jacanas, little grassbird, crakes, rails, swamp harrier, yellow chat	None	stilts, avocets
Refuge area for:					
Kingfisher, certain ducks	Snipe, plovers	Little grassbird, yellow chat	Waterhens, ducks, grebes, bitterns, little grassbird, crakes, rails, reed warbler, yellow chat	Ducks, swans, grebes, gulls, terns	Ducks, swans, gulls, terns

and creeks have been omitted as a separate type for they scarcely rate as a *major* waterbird habitat over most of Australia. Considerable seasonal fluctuations in bird numbers occur in several of these habitat types. Mostly in southeastern Australia, waterbird numbers are maximal in summer–autumn, but elsewhere maximum numbers generally occur whenever rainfall patterns produce suitable wetland habitats.

Waterbird habitats have also been classified according to their salinity, depth and position in relation to the coast. Such a classification was derived basically for conservation and management purposes and is less useful for present purposes than that indicated in Table 7.2.

8 Mammals

Four mammals need consideration, three native and one exotic. The natives are the platypus, the water rat and the false water rat; the exotic is the water buffalo. Though the dugong or sea-cow (order Sirenia) occurs in Australia, it is entirely marine in distribution, unlike its northern and only living relative, the manatee, found in both fresh and marine waters.

Fig. 8.1 *Platypus.* (a) Close view of head to show bill and eyes; (b) whole specimen. Photographs by courtesy of Dr P. D. Temple-Smith.

PLATYPUS

The platypus (*Ornithorhynchus anatinus*) (Fig. 8.1(a) and (b)) is so unique a mammal that — together with the echidna — it is placed in a subclass, the Prototheria, distinct from the subclass Theria, which contains all other living mammals. It is found only in Australia. In the circumstances, a slightly overlong account is appropriate.

Platypus anatomy is a curious mixture of mammalian features, reptilian characters, and characters unique to the genus. The skeleton best displays the reptilian relationships: various anatomical details of the skull, the arrangement of the ear bones, the nature of the vertebrae (particularly the cervical), and the structure of the pectoral and pelvic girdles and attached limbs all show reptilian features. Non-skeletal evidence of reptilian affinities includes the presence of a common cloaca into which both the rectum and urinogenital systems open, and the persistence of the egg-laying habit. Despite these features, however, the platypus is unequivocally a mammal as indicated by other anatomical features of the skull, the presence of fur, the structure of the brain, warm-bloodedness, the habit of suckling its young with milk from specialized sweat glands, the presence of a diaphragm separating thoracic and abdominal cavities, and by the structure of its heart with its single left aortic arch. The unique features of the platypus can mostly be regarded as adaptations to its specialized mode of life: the duck-like bill, dorsal nostrils, long palate, paddle-like tail, webbed feet, and lack of external ears.

Not surprisingly, a duck-billed, web-footed, fur-coated animal when first discovered was a curiosity, and a brief recapitulation of the original discovery and subsequent events is of interest. The platypus was apparently first seen by the English colonists in 1797 when it was observed on the banks of the Hawkesbury River, New South Wales, and reported as an 'amphibious animal of the mole species'. The first written account of its natural occurrence was in a work by David Collins published in 1802, but a specimen was described three years earlier (1799) by George Shaw of the British Museum based on material sent to England in 1798. Shaw used the name *Platypus anatinus*, but *Platypus* was subsequently shown to be preoccupied and the generic name was changed to *Ornithorhynchus*. However, platypus gradually became the vernacular name and has replaced other early ones such as duckbill, duckmole and water-mole. Shaw suspected some deception and there is no doubt that his examination was very critical; of the beak, for example, he noted:

> 'nor is it without the most minute and rigid examination that we can per-
> suade ourselves of its being the real beak or snout of a quadruped . . . on a
> subject so extraordinary as the present, a degree of scepticism is not only
> pardonable, but laudable; and I ought perhaps to acknowledge that I
> almost doubt the testimony of my own eyes with respect of the structure of
> this animal's beak; yet must confess that I can perceive no appearance of
> any deceptive preparation . . . nor can the most accurate examination of
> expert anatomists discover any deception in this particular.'

By 1802, and following dissection of a specimen by Sir Everard Home, a leading anatomist, the scientific world had accepted the authenticity of the animal. The question then to be resolved was its position in the Vertebrata: was it a mammal? Geoffroy, the French naturalist, believed not, whereas English naturalists thought otherwise. The question was eventually resolved, especially when, in 1824, the German anatomist Meckel discovered mammary glands. Perhaps fortunately at that time, it was not known whether eggs were laid, although the possibility was recognized by the early anatomists. The egg-laying habit was not confirmed until as late as 1884 when W. H. Caldwell, an

English zoologist who came to Australia specifically to study reproduction in native mammals, was able to authenticate the habit. It was announced dramatically in a cable sent to the British Association for the Advancement of Science, then meeting in Montreal. Mention should be made, however, of a short note published in the first volume of the *Victorian Naturalist* (in 1884) by a Mr F. J. Williams and based on a paper read in October 1880. This note states that:

> 'My experience teaches me that the Platypus builds a warm and secure nest fit to hold eggs, or rear the young ones in. It has eggs inside in November, in some localities it may be earlier, or later, and its body contains milk fit to nourish the young.'

Unfortunately it is not possible to decide unequivocally whether the author meant that the animal had eggs inside the nest or inside its body.

When discovered and for several decades afterwards, the platypus was common throughout its range; it was certainly numerous in the Blue Mountains, for example, when Governor Macquarie journeyed across them in 1815. Unfortunately for the platypus, it had a fur pelt; the result was wholesale and remorseless slaughter. It is not too surprising then that by 1852 the platypus was reported as extinct in the Blue Mountains and rare elsewhere. By the beginning of this century, all Australian states where it occurred recognized the need for its protection, and appropriate legislation became law. The platypus is now a fully protected animal.

No doubt the induced rarity of the species and its secretive and crepuscular habits explain in part why so little is known of its ecology. But it is nonetheless surprising that we do know so comparatively little of a precise sort on this subject. Until recently, indeed, much of our knowledge of platypus natural history and behaviour was largely anecdotal; and a fair proportion also was derived from observations of a few captive specimens. The following account is based mostly on recent field and laboratory investigations.

The natural distribution of the platypus is very extensive; it occurs in most of the large rivers and streams east and west of the Great Dividing Range from Cape York in the north of Queensland to southern Victoria. It also occurs in Tasmania and South Australia, though in the latter state it was largely exterminated and recent rare sightings on the Murray may represent a reinvasion from the east. It has been introduced to Kangaroo Island in South Australia where a viable population maintains itself in the Flinders Chase national park. It is not known from Papua New Guinea, from Western Australia, nor from western New South Wales — the Darling and its tributaries appear never to have supported populations. Until recently, its contemporary occurrence was thought to be rare, but evidence now shows that the rigid protection offered the species has had an effect, and the species today is reasonably common.

With regard to the nature of its habitat, the platypus does not seem overly restrictive: muddy lowland rivers, upland streams, highland lakes and most of the larger man-made impoundments are all inhabited. However, discontinuity of river flows seems to discourage its occurrence and heat stress occurs at air temperatures above 25°C. Throughout the extensive geographical and habitat range, only one species seems involved, and the status of various described subspecies (races) requires validation.

Although the platypus appears now to be reasonably common, few people see them in the wild. This is because the animal is generally inactive during the day; most daylight hours are spent sleeping in a burrow situated in the bank well above water level and some 2 m from the water's edge. The active periods are dawn and dusk. At these times, individuals enter water where they forage for food using a very sensitive bill, but not their tightly shut eyes and ears. The submerged platypus is a graceful and adept swimmer.

Food comprises a variety of aquatic invertebrates including mussels, insects, crustaceans and oligochaetes probably supplemented by various terrestrial insects. The platypus is an opportunistic feeder. There is undoubtedly a wide trophic overlap with several other aquatic vertebrates — tortoises, trout, various native fish — but this appears to be of no great significance: the species seems quite capable of holding its own in this competitive situation. One of the major difficulties in keeping captive specimens relates to the vast amount of food that needs to be provided.

Known predators are few. Some birds of prey, some of the larger carnivorous marsupials, certain snakes and varanid lizards, and the Murray cod appear to be natural predators, with foxes and feral cats introduced ones. The main danger to the platypus remains man; although direct exploitation has ceased, a number of man's activities is inimical to the species. The trampling of burrows by grazing stock and the drowning of feeding individuals caught in fish nets are two results of man's activities. Competition from the introduced rabbit for suitable burrowing space on stretches of river bank may be another pressure.

Reproduction is a seasonal phenomenon. Both males and females show distinct seasonal changes in such features as body weight, tail fat reserves and moulting that correlate with seasonal changes in the sexual organs. In the male, the testes begin to develop in May in southeastern Australia, and males are sexually most active during spring. The crural gland is located at the base of the hind limb and discharges a strong poison via a pair of large spurs on the hind feet, and it also exhibits seasonal changes (the gland and spurs are absent in females). The function of the poison spurs seems to be to ensure territorial separation of males; apparently their main function is as a weapon of offence in combat with other males. Male spurs should be avoided if handling specimens, for the poison they inject, though not lethal to man, causes an extremely painful reaction. In females, from one to three soft-shelled, yolk-laden eggs are incubated on the belly during confinement within a nest sealed inside a special breeding burrow. This occurs in November and December in southeast Australia. Incubation lasts for about 10 days. The young are suckled as other mammals, but not from nipples; they simply lick up milk exuded from mammary glands. The young spend some 4 months in the nest before weaning is complete.

NATIVE RATS

Both rats are true rodents and therefore placental mammals, not marsupials. They belong in the family Muridae, and in this to the subfamily Hydromyinae, a group of rodents that shows greatest diversity in Papua New Guinea but which has two genera in Australia. One of these (*Hydromys*) is common to Papua New Guinea and Australia; the other (*Xeromys*) is endemic to Australia. The Hydromyinae are specialized for a subaquatic or aquatic existence and apparently originated and underwent adaptive evolution in Papua New Guinea. From there, it has been suggested, one lineage (*Xeromys*) reached Australia early in the Miocene (some 26 million years ago), and another (within *Hydromys*) later in the same period. Other, more recent, evidence suggests that the rodent invasion of Australia did not occur until much later than the Miocene — until at least the Pliocene. Irrespective of the time of arrival, ancestors of the native water rats were probably amongst the first murid rodents to reach Australia. Whether they then replaced any marsupial forms occupying aquatic habitats is not known, but certainly their arrival on the Australian mainland was after the major marsupial adaptive radiation of the early Tertiary. Whatever the case, Australia does not have a marsupial which parallels placental adaptations to a life spent associated with fresh waters, as exemplified by the placental beaver or otter. The only Australian mammals approaching such adaptations are placentals too. Strangely, South America does have an aquatic marsupial (*Chironectes*).

The best adapted of the two rats is the so-called water rat or beaver rat (*Hydromys chrysogaster*) (Fig. 8.2). It is widely distributed within Australia (Fig. 8.3(a)), and also occurs in Papua New Guinea. At first numerous, populations declined in certain areas following its exploitation for fur. Destruction as vermin and predation from feral cats also probably contributed to the decline. Recent legislation now protects the species in all states (except for a short season in Tasmania), and populations have recovered. The species can today be regarded as widespread and common. The fur itelf varies in col-

Fig. 8.2 The water rat. Photograph by courtesy of Dr A. C. Robinson.

our; in the southwest it is black, in the southeast (including Tasmania) golden-brown and orange to grey and white, and in the north, greyish-brown.

The habitat of the water rat is the banks of rivers and lakes, but it is not restricted narrowly to these; it also occurs on marine shores, in estuaries, and on several offshore islands. Moreover, considerable overland journeys may be undertaken. It may reach a length of some 30 cm (with an additional 25 cm of tail), and has a set of extremely sharp, chisel-like upper incisors and pointed lower teeth, proving a formidable captive if provoked. Under normal cir-

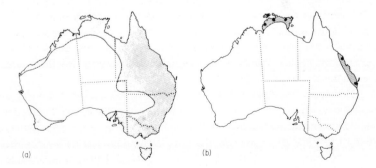

Fig. 8.3 Geographical distribution of (a) water rat and (b) false water rat. In (b), all known localities are plotted individually. Redrawn and modified from Watts and Aslin (1981).

cumstances, however, the animal is shy and timid. Large adults may reach a weight of 1 kg and the species is therefore one of the largest of native rodent species. Adaptations to an aquatic life include a water-repelling fur, and broad and webbed, paddle-like hind feet. It is partly diurnal in habits, but days are usually spent in bankside burrows, rock crevices, debris, and other suitable refuges. Such behaviour may be related to the fact that the species is prone to heat stress. In fact, overall, it has poor thermoregulatory abilities and apparently compensates for this behaviourally. The burrowing activities of large populations may cause considerable damage to channel banks and other water-control structures in irrigation areas.

Mainly, but not entirely carnivorous, the water rat usually feeds on large aquatic insects, yabbies, mussels, fish, frogs, tortoises and small warm-blooded vertebrates. It appears to be one of the few native predators of the introduced cane toad. Often it has a favourite food platform. Snakes and various birds of prey are its natural predators, but feral cats also feed upon it.

Breeding biology has been studied in northern Victoria and mid-New South Wales. In both regions, breeding occurs mainly in spring and summer. However, populations in central regions of Australia probably breed whenever suitable conditions occur. Litter size varies from one to seven, but is usually four or five. Although growth is rapid, both sexes probably do not mature sexually until more than 1 year old. Apparently one litter per season is normal, but under favourable conditions it is thought that up to three may be produced.

Much less is known about the biology of the other native water rat, the rare false water rat or false swamp rat (*Xeromys myoides*). This species is smaller, with large adults weighing less than 50 g, and it does not have webbed feet. It does, however, as an obvious adaptation to its aquatic habitat, have a grey water-repelling fur. Isolated areas in eastern Queensland and the Northern Territory represent its known area of distribution (Fig. 8.3(b)), but it may be more widespread and common than present records indicate. Recent observations suggest that it is an active climber and rather less aquatic in habits than was previously thought. Brackish (mangrove) as well as freshwater areas are inhabited.

Perhaps a few other rodents that are not strictly aquatic but which are nevertheless characteristically associated with swamps, lakes and watercourses should also be mentioned here. The swamp rat (*Rattus lutreolus*) is mainly a southeastern species of swamps and reedy areas along coastal watercourses. The more restricted dusky rat (*R. colletti*) is confined to the coastal floodplain and swamps of the most northerly part of the Northern Territory. Yet other species may be found in swampy areas in parts of their range and amongst drier vegetation elsewhere. The broad-toothed rat (*Mastacomys fuscus*) is found in swamps in Tasmania, but in drier, forested areas in Victoria. Tunney's rat (*R. tunneyi*) occurs near watercourses in the Northern Territory and in swamps on offshore islands of Queensland, but in much drier habitats in mainland Queensland. And in parts of southwestern Western Australia, the bush rat (*R. fuscipes*) inhabits wet streamside vegetation and swamps, whereas elsewhere it inhabits much drier areas.

WATER BUFFALO

The water buffalo (*Bubalus bubalis*), sometimes called the Asian buffalo, was introduced to northern Australia between 1825 and 1829 (Fig. 8.4). Since then is has proliferated and now occurs widely as a feral and semi-domesticated species over the northern swamp lands of the Northern Territory. Although only one calf is produced per year, recent population estimates range up to 200 000. For the most part, animals are found wading in or near freshwater lagoons, swamps, and rivers, frequently swimming and wallowing in the mud of these habitats. Feeding is usually a morning or late afternoon activity, and food is a wide selection of marsh vegetation and hydrophytes. Stomach sampl-

Fig. 8.4 The water buffalo. A bull emerging from Cannon Hill Lagoon, East Alligator river region. Photograph by courtesy of Mr G. Miles.

Fig. 8.5 Environmental damage caused by water buffalo. (a) An area on Magela Creek floodplain where wallows are extensive, paperbark vegetation (*Melaleuca leucadendron*) has been killed, and where there is now no tree recruitment; (b) an area in the Kakadu National Park showing edge of billabong now denuded of vegetation (with resultant sheet erosion) and where all seedling trees are eaten. Photographs by courtesy of Mr G. Miles.

ing has indicated that, when hungry, buffaloes will eat the bark of trees, paperbark leaves, twigs, and other 'unpalatable' material. Where large populations occur, overgrazing has been reported.

The main mating season occurs during the wet months (October to about May), when males congregate with females and the young are born. Bulls and cows inhabit separate areas during the dry season (May to about September). As a general rule, strong site attachment is established and herds (50–500 animals) have restricted home ranges.

There are both advantages and disadvantages attached to the presence of the buffalo in Australia. Disadvantages are that herds often overgraze, destroy the natural character and vegetation of many water-bodies by excessive wallowing (Fig. 8.5), and generally modify unfavourably the environment of the northern black soil plains. On the other hand, not noticeably more prone to disease than domesticated cattle, buffaloes are a source of meat and hides, and return an annual income in excess of $1 million for meat alone. They are also something of a tourist attraction, being large powerful animals, some 3 m long, with large, curved horns. Despite their somewhat prepossessing appearance and reputation, they are shy animals, and under normal circumstances are difficult to approach closely. For a variety of reasons, but particularly because of environmental damage, their closer management is called for.

BUNYIPS

It is possible that yet another mammal occurred in Australian fresh waters, for there have been many reports of a hairy aquatic animal clearly neither water rat nor platypus. Considerable mystery surrounds this animal and it has come to be called the bunyip, a word of Aboriginal derivation roughly equivalent in meaning to 'devil' or 'spirit'. Lacking specimens, there is no evidence to decide unequivocally as to whether it was a mammal, or indeed, if it existed at all.

Perhaps the first report by white man was the mysterious roaring heard by the crew of the *Géographe* in 1801 in the Swan River, Western Australia. The first sighting by Europeans would seem to have been that by the explorer Hamilton Hume in 1821; he reported the occurrence of an animal not unlike a manatee or hippopotamus in Lake Bathurst, New South Wales. No specimens were collected, but for several years afterwards there were reports of mysterious animals near Bathurst. Then, in 1846 a skull purported by Aborigines to be that of a bunyip was found on the banks of the River Murrumbidgee and sent to W. S. McLeay in Sydney. From there an illustration was sent to London and examined by Professor Sir Richard Owen. He decided that it was a calf's skull. This skull is now in the McLeay Museum at the University of Sydney (Fig. 8.6). It is said to be probably the malformed skull of a calf or ox.

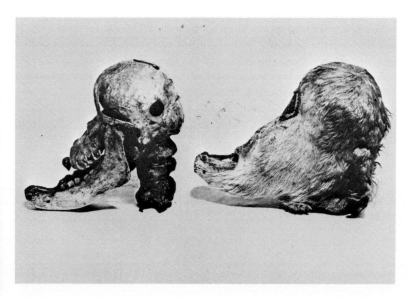

Fig. 8.6 Skull of 'bunyip' collected from the banks of the River Murrumbidgee in 1846 and in possession of the Macleay Museum, University of Sydney. This is the skull long thought to have disappeared by many authors. It was relocated by Dr P. J. Stanbury who also took the photograph.

Notwithstanding the opinion of Richard Owen, from 1846 on many bunyip sightings were made, particularly in New South Wales, Victoria and Tasmania. A remarkable feature of these was their degree of agreement. Thus, the bunyip reported from Lake Tiberias, Tasmania, in 1852 had a bull-dog head, short legs and was hairy. An animal like a sheep-dog with two small flippers was reported in 1863 from the Great Lake, Tasmania. In Lake Burrumbeet, Victoria, an aquatic animal like a big retriever dog with a round head and hardly any ears was sighted in 1872. And from Dalby, Queensland, there was a report in 1873 of a creature not unlike a seal. Twentieth century reports are less frequent and credible so that if there ever was a bunyip in Australia it would now appear extinct or virtually so.

Allowing at least some credence to the early reports, the question, then, is what sort of animal was the bunyip? Two possible answers have been proposed. The bunyip could have been an extinct marsupial adapted to aquatic life in the way that some modern placentals are (e.g. the otter) and knowledge of which was retained in Aboriginal folklore. Or bunyip reports could have been sightings of seals that had wandered inland. This is the most plausible answer for several seals have been reported and some actually caught well inland; there are old records of seals having penetrated over 1000 km upstream in the River Murray, for example. The presence of locks and weirs on this river would now make such penetration impossible, and account for the 'extinction' of the bunyip in at least this river. It is possible of course that in the early days of European settlement, cattle formed the basis of Aboriginal stories about bunyips: to a people knowing only kangaroos as large mammals, long-horned English cows must have appeared strange indeed.

9 Microscopic plants

Many plants or plant-like organisms in inland waters are microscopic, although their mass in any given water-body may greatly exceed that of macrophytes. In traditional terms, four types are involved: bacteria, fungi, Cyanobacteria (blue-green algae), and true algae. Modern systems of classifying living organisms often place them into three major groups: bacteria and Cyanobacteria (Kingdom Monera), fungi (Fungi), algae (Plantae). Older systems grouped them all as plants. The size range of individuals is extensive. Bacteria are the smallest, with lengths from less than 0.2 μm up to 50 μm, whereas some algae are visible to the naked eye. Though microscopic, individuals are often so numerous *en masse* that populations are clearly evident, as in salt lakes when immense numbers of bacteria impart a red coloration to the water, or in fresh lakes when cyanobacterial blooms form thick surface scums. Whatever their mass, the activities of microscopic plants or plant-like organisms within aquatic ecosystems are essential to the functioning of these.

This chapter is mostly about the Cyanobacteria and algae. This is not because bacteria and fungi are unimportant in aquatic environments, but because the Cyanobacteria and algae are the most obvious elements of the microscopic 'flora' and we know more about them. Most of the attention given to Australian aquatic bacteria has centred on the presence of human pathogens which under natural circumstances should not occur in inland waters. Although a few endemic algal species are known, most species of algae and Cyanobacteria in Australia appear to be cosmopolitan or ubiquitous. Additionally, a few algae are common elsewhere but not in Australia (e.g. *Stephanodiscus*). This chapter, then, unlike others dealing with aquatic organisms (Chapters 4–8, 10), is concerned more with providing information of general interest than with highlighting the special features of Australian forms.

BACTERIA

Although, in principle, bacteria are classified according to the rules originally formulated by biologists for larger organisms, it is doubtful if bacterial 'species' equate to plant or animal species. Similar difficulties occur in their higher classification (to orders, families). Thus, the use of scientific names for bacteria in aquatic environments is often a matter of convenience and a recognition of an ecological relationship, not a firm taxonomic statement. Whatever the case, the ecological and taxonomic diversity of aquatic bacteria is very great; they occupy a multiplicity of ecological niches and occur in freshwater lakes and streams, highly saline pans and hot springs. They come from most of the formally recognized families of bacteria and all but one of the orders.

Of particular interest in Australia, a continent with many salt lakes, are those in highly saline waters, the halophilic bacteria. The two main genera described thus far are *Halobacterium* and *Halococcus*, both organisms of the brine rather than the sediments. Their adaptations to life in saline water include high internal concentrations of sodium, potassium and chloride as a counter to the high external salinity, and, it appears, the development of special red, pink or orange pigments to protect against high light intensities. Another interesting feature is that at least one species, *Halobacterium halobium*, has the unique ability to harness light energy in anaerobic conditions without chlorophyll. It uses a special purple pigment, bacteriorhodopsin (a carotenoid), located in its plasma membrane. The process does not involve the fixation of carbon, and its adaptive value within salt lakes remains unclear. Of other bacteria, the most readily visible are photosynthetic green and purple sulphur bacteria. They occur in anaerobic regions where hydrogen sulphide is abundant.

Within a water-body, bacteria are found in most microhabitats. They occur in the water column of lakes, reservoirs and other standing bodies of water, where they are commonly associated with organic detritus or with other microorganisms. They occur in even greater numbers in lake sediments. They are important in the epiphytic microbial community coating the submerged surfaces of macrophytes, and they are found attached to rocks and other bottom materials in rivers and streams. Where running waters are organically polluted, bacteria are particularly numerous (see Fig. 12.2), and they are a major component of sewage 'fungus', large, clearly visible, and characteristic ragged masses of white, yellow or brownish material. *Sphaerotilus natans* and *Beggiatoa alba* are important bacterial members of the sewage 'fungus' community.

The functional roles of bacteria within aquatic ecosystems are important and diverse. Most of the bacteria are aerobic heterotrophs, requiring organic substances as a source of energy in the presence of oxygen. The energy source is largely dead plant or animal tissue (or faeces). Most, then, are saprophytes, and they are agents preventing the accumulation of dead plants or animals.

Almost all types of organic material are used, but individual bacterial requirements are often specific. Some heterotrophic bacteria can use only a limited number of sugars, others feed on chitin, and yet others utilize dissolved organic materials such as glycollic acid secreted into the water by other plants. Fresh terrestrial plant litter that has fallen into water is apparently not an important source of bacterial food and the role of bacteria in the initial stages of litter decomposition seems to be relatively unimportant.

Because most bacterial breakdown of organic material is aerobic, there are important physicochemical repercussions when large numbers of bacteria build up. In the water column of lakes, oxygen concentrations may show a sharp vertical decrease, and the lower water layers may come to lack oxygen altogether. In the sediments, likewise, there is a vertical microzonation, with only the upper layers oxygenated, or, if all oxygen is used up at the sediment surface and not replaced by diffusion from overlying water, the sediments may become entirely anoxic. And in streams and rivers, there is often a 'sag' in the concentration of oxygen downstream of a point source of organic pollution (the oxygen 'sag' curve; Fig. 12.2).

Whilst most types of aquatic bacteria are aerobic heterotrophs, other sorts of metabolic pathways are utilized. In anoxic sediments and waters, for example, there are bacteria which obtain energy either photosynthetically or chemosynthetically to synthesize organic matter from carbon dioxide. Examples of anaerobic photosynthetic bacteria were given in the discussion of salt lake bacteria. Examples of chemosynthetic bacteria are those obtaining energy initially from chemical transformations of various nitrogen, sulphur and iron compounds. However, with the exception of denitrifying bacteria (most of which, in any case, are heterotrophs), chemosynthetic bacteria are generally regarded as unimportant in aquatic ecosystems.

Other bacterial groups include the methanogenic bacteria which convert carbon dioxide to methane under anoxic conditions, the methane-oxidizing bacteria which do the reverse under oxygenated conditions, bacteria of the sulphur cycle (both as reducing and oxidizing agents), and bacterial reducers of iron. The bacterial conversion of inorganic mercury to highly toxic organic mercury (chiefly methyl mercury) is significant for the organic form of mercury may accumulate in the food chain. Finally, *Hyphomicrobium*, a budding bacterium, brings about the deposition of manganese in pipelines used to transport water, causing an effective loss of carrying capacity. In Tasmania, it has cost the Hydro Electric Commission (by 1973) some $3 million.

FUNGI

Most aquatic fungi are placed in the composite fungal class, the

'Phycomycetes', but there are also many in the higher fungal classes (Ascomycetes, Basidiomycetes, Fungi Imperfecti). All are incapable of manufacturing their own food from carbon dioxide, as plants can, and require an organic food source. This comprises a variety of dead plants and animals for saprophytic fungi, or live organisms such as algae or fish for parasitic fungi. The structure of aquatic fungi varies from unicellular to a complex mass of filaments. A few, such as the yeasts, grow freely in water, but the majority is associated with submerged surfaces. Mostly aerobic, they occur widely in freshwater lakes and ponds, rivers and streams, but not in markedly saline waters. Organically polluted fresh waters are particularly favoured habitats.

Although the broad outlines of the ecology of aquatic fungi have been known for some time, it is only recently that we have arrived at a more precise awareness of their roles within aquatic ecosystems. It now appears, for example, that they play an important role in the breakdown of macrophyte litter. Most work on this subject concerns the fungal breakdown of deciduous leaves and has taken place in the northern hemisphere. The nature of the process and the fungal succession involved in the breakdown of non-deciduous litter in Australian fresh waters are undoubtedly different. Nevertheless, the occurrence in Australia of those fungi typically involved in litter breakdown (Hyphomycetes) has long been recorded.

CYANOBACTERIA AND ALGAE

Composition and distribution

For many years, two groups of (largely) microscopic organisms were referred to as 'algae'. However, though rather similar in gross external appearance, they are not related. One group, the Cyanobacteria or so-called blue-green algae, is allied to the bacteria and has procaryotic cells. Procaryotic cells lack a nuclear membrane and many intracellular structures found in eucaryotic cells. The second group, the algae proper, has eucaryotic cells. It is convenient to treat both Cyanobacteria and algae together.

The Cyanobacteria are relatively homogeneous in basic structure, but vary in form from minute, single-celled organisms to filamentous and colonial ones. Unlike algae, some can obtain their nitrogen requirements from dissolved nitrogen. An exceedingly ancient group of photosynthetic microorganisms, most species of Cyanobacteria are cosmopolitan. A few important genera are illustrated in Fig. 9.1. Several species are nuisance organisms, as will be discussed later.

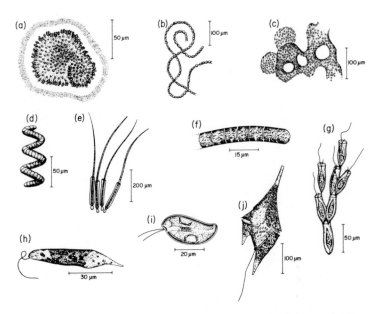

Fig. 9.1 Various genera of Cyanobacteria ((a)–(f)) and algae ((g)–(j)). (a) *Gomphosphaeria* colony; (b) *Anabaena*; (c) *Microcystis* colony; (d) *Spirulina*; (e) *Gloeotrichia*; (f) *Oscillatoria*; (g) *Dinobryon* colony (Chrysophyta); (h) *Euglena* (Euglenophyta); (i) *Cryptomonas* (Cryptophyta); (j) *Ceratium* (Dinophyta). Mainly redrawn after Belcher and Swale (1976).

The algae are more diverse. Twelve major subgroups (phyla) are commonly accepted, of which one, the Charophyta or stoneworts, comprises large plants and is discussed in the next chapter. The rest — or at least those in fresh waters — are mostly microscopic. All algal phyla have at least some representatives in fresh waters, but two phyla are mainly marine. Of the typically freshwater phyla, the most important are the green algae (Chlorophyta) including the desmids, the diatoms (Bacillariophyta), the dinoflagellates (Dinophyta), the Chrysophyta, and the euglenoids (Euglenophyta). Others may be locally or occasionally important. Most species appear to be ubiquitous or cosmopolitan, but since algal systematics are presently in chaos (according to many algal taxonomists), the complete truth of that statement awaits verification. In any event, several endemic Australian species have been described, of which perhaps the best-known is *Micrasterias hardyi* (Fig. 9.2(i)), recognized first from the Yan Yean Reservoir near Melbourne. Representative genera are illustrated in Figs 9.1 to 9.3.

Figures 9.1 to 9.3 indicate the wide range in size and shape of Cyanobacteria and algae. Some are minute with individual cells of a size less

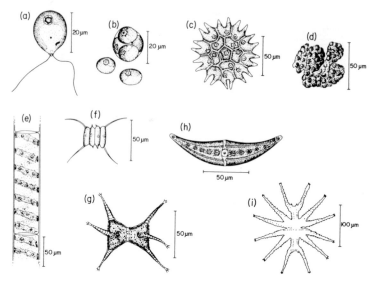

Fig. 9.2 Various genera of green algae (Chlorophyta) including desmid genera ((g)–(i)). (a) *Chlamydomonas*; (b) *Chlorella* colony and single cells; (c) *Pediastrum* colony; (d) *Botryococcus braunii*; (e) part of *Spirogyra* filament; (f) *Scenedesmus* colony; (g) *Staurastrum*; (h) *Closterium*; (i) *Micrasterias hardyi*. Mainly redrawn after Belcher and Swale (1976).

than 5 μm. Others (not illustrated) are visible to the naked eye. Shapes vary from a simple globular form to a highly complex or filamentous one, and individual cells may lead an isolated existence, or associate with others to form large colonies.

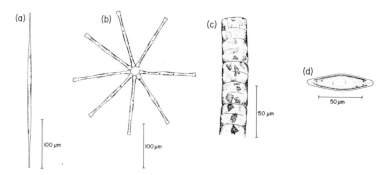

Fig. 9.3 Some diatom genera (Bacillariophyta). (a) *Synedra*; (b) *Asterionella*; (c) *Melosira*; (d) *Navicula*. All redrawn after Belcher and Swale (1979).

All types of inland surface water contain Cyanobacteria and algae, although in certain waters diversity and numbers may be depressed. Cyanobacteria are often the only forms found in hot springs, and *Dunaliella*, a green alga, is often the only alga present in highly saline lakes. Special adaptations are a feature in such physiologically stressful environments; for *Dunaliella*, an important one is the presence of large amounts of intracellular glycerol which preserves osmotic balance. Together with the halobacteria, *Dunaliella* causes the red colour of many salt lakes.

Ecological considerations

Three major sorts of habitat may be distinguished. First, there is the water column of standing water-bodies such as lakes, reservoirs and ponds. The Cyanobacteria and algae of this habitat are collectively termed phytoplankton, and occur mostly in the upper layers (depending on circulation patterns). A second major sort involves bottom and marginal material: littoral sediments, stones in streams, and similar substrates are the main sites. And third, the submerged surfaces of macrophytes provide important sites for Cyanobacteria and algae.

Most phytoplankton species are denser than water and hence tend to sink. Exceptions are certain Cyanobacteria which have gas vacuoles making them lighter than water, and *Botryococcus braunii*, a green alga which may be rich in hydrocarbons.

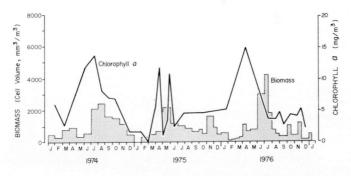

Fig. 9.4 Phytoplankton biomass and chlorophyll *a* concentration in Lake Hume, 1974–76. Chlorophyll data refer to surface samples; biomass data are integrated values for the uppermost 3 m stratum, except during August–December 1975, when only surface samples were taken. From Walker and Hillman (1977).

Marked fluctuations over time in numbers (or biomass), species composition, and photosynthetic activity are typical features of phytoplankton
populations. Figures 9.4 to 9.6 illustrate this for Lake Hume, a large impoundment of the River Murray near Albury. Figure 9.4 shows the seasonal variation
in biomass as total cell volume in a standard volume of water and as
chlorophyll *a* concentration (a readily measured and reasonably accurate index
of phytoplankton biomass). By whatever measure, large fluctuations in
biomass are clearly evident between both seasons and years. In many lakes,
phytoplankton biomass typically exhibits a pattern with peak values in
spring–early summer, and a smaller peak in autumn. However, this pattern
did not prevail in Lake Hume during the years 1974–76 (Fig. 9.4). Clearly,
considerable differences *between* lakes represent yet another aspect of
phytoplankton variability.

The way in which phytoplankton composition varies over time in the
same lake is shown in Fig. 9.5 (Lake Hume). Again, marked fluctuations are
evident; whilst diatoms (species of *Melosira*) are almost always dominant,
other algae and Cyanobacteria (*Anabaena spiroides*) are sometimes important. Again, different patterns may occur elsewhere. In Mount Bold Reservoir,
near Adelaide, for example, diatoms (chiefly *Cyclotella*) dominate only in
winter, with green algae important in late spring, Cyanobacteria (*Anabaena,
Gomphosphaeria, Microcystis*) then motile algae in summer to early autumn,
and green algae in autumn. Variations in the intensity of photosynthetic activity with time and at different depths are shown in Figs 9.6 and 9.7 (Lake
Hume). Horizontal variation is also a feature, particularly in near-surface
waters.

All in all, phytoplankton populations, together with their associated
zooplankton populations, form a complex community whose dynamics are far
from being understood. For even the simplest of correlations, for example that
between biomass and nutrient concentrations, our predictive ability is still

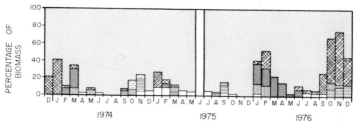

Fig. 9.5   Relative composition of phytoplankton populations at one station in
Lake Hume, 1973–76. After Walker and Hillman (1977).   Bacillariophyta;
Chrysophyta; Chlorophyta; Cyanobacteria; Euglenophyta;
Cryptophyta; unidentified.

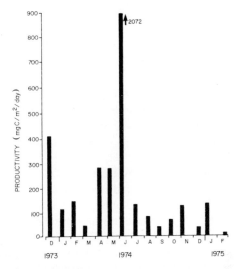

Fig. 9.6 Day rates of phytoplankton productivity in Lake Hume, 1974. From Walker and Hillman (1977).

small. We are, nevertheless, slowly disentangling the web of interrelated determinands controlling phytoplankton distribution and abundance; we are becoming aware of how phytoplankton growth, productivity and composition are determined by the light regime, inorganic nutrient concentrations, temperature and water turbulence (as principal determinands). However, we are far from understanding the role of organic nutrients in phytoplankton dynamics, though it may be important since most algae are unable to manufacture at least one, but often more, essential organic compounds (e.g. certain vitamins).

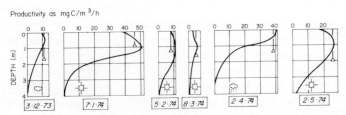

Fig. 9.7 Areal day rates of phytoplankton productivity at one station in Lake Hume. Note: Secchi transparency is the depth at which a circular disc, 20 cm diameter and painted black and white, can just be seen. From Walker and Hillman (1977). Secchi transparency (d). The bottom of the symbol indicates the depth of the Secchi transparency. weather symbols.

The Cyanobacteria and algae associated with littoral sediments and other submerged, non-living material constitute a quite different, much less known community. Two major groups can be distinguished, that on mud or sand, and that on rock. Not surprisingly, many different and distinct communities within both groups have been found according to the particular nature of the surface and type of water-body involved. Both motile and attached forms are associated with mud or sand, and diatoms are important. Those associated with rock surfaces frequently include encrusting or attached filamentous Cyanobacteria, green and red algae, as well as encrusting diatoms. Relatively pure stands of species often occur.

Forms associated with submerged macrophyte surfaces give rise to more complex communities (although again diatoms are particularly common) that may include bacteria, protozoa and other microorganisms. They probably have a greater need of preformed organic material than those in the phytoplankton or associated with non-living submerged surfaces. Many submerged macrophytes have been shown actively to secrete a variety of soluble organic substances; many of these can be readily absorbed by at least some associated algae and bacteria.

Economic significance

Cyanobacteria and algae are often of considerable economic significance. They have many direct uses, although their presence in water may cause several problems.

A major problem exists when too much cyanobacterial or algal biomass occurs in water used for drinking or industrial supplies. Certain diatoms (*Melosira, Asterionella, Synedra*) are common nuisance organisms which may clog screens and filters in the water treatment works. Excessive biomass also causes problems when it results in high levels of colour or turbidity, when it decays in distribution systems to give rise to perhaps further discoloration and to deoxygenation, and when it promotes animal growths which may require additional chlorination. Cyanobacteria (e.g. *Anabaena*) or algae may give rise to odour and taste problems, may stain clothes being washed, or significantly impair water use for industries with high water quality requirements. Fluctuations in biomass alone are a problem (to engineers) in that variable water quality is an undesirable feature of a water supply—whatever the average quality.

Excessive biomass gives rise to another set of problems in lakes and reservoirs not used to supply water. They are largely aesthetic ones; unsightly masses may smell unpleasant, and the formation of thick scums often detracts from the pleasure of swimming and boating. Lake Burley Griffin has incipient problems of this type.

The presence of excess biomass is usually a feature of the process of 'eutrophication', the enrichment of inland waters with plant nutrients (of which phosphorus and nitrogen are the most significant). For the most part, the enrichment reflects man's activities; the nutrients arrive in sewage or in run-off from altered catchments. Cyanobacteria are often important in enriched lakes, and because many of them float they may accumulate in vast numbers near the surface. When they do this they are said to 'bloom'. Apart from their unaesthetic attributes, blooms are often undesirable because many Cyanobacteria produce toxins. As long ago as 1878, stock deaths were reported in cattle that drank water containing Cyanobacteria from Lake Alexandrina in South Australia, and there have been many similar reports since then. Likewise, cyanobacterial toxins have been implicated in fish kills, and at least once in medical problems such as gastroenteritis.

Control of cyanobacterial and algal biomass in water supply reservoirs is often by regular dosing with copper sulphate or other algicides, but this, whilst usually effective, is becoming increasingly expensive, and is of course only a temporary solution and may affect other components of the food chain. More thought is now being given to control measures that manipulate the environment so as to make it less favourable for plant growth. Artificial aeration which increases the depth of the layer inhabited by Cyanobacteria and algae (the mixing zone) relative to the illuminated layer is one such method. Long-term control of biomass in lakes artificially enriched by sewage addition, however, can only be brought about by sewage diversion or by otherwise preventing the addition of plant nutrients in sewage and other inflowing waters. There is already a number of cases where sewage has been diverted away from enriched lakes with a subsequent decrease in biomass (though recovery rates are variable). Lake Washington in the United States is perhaps the best known example. There is none in Australia.

Other means of control (or of lake restoration) are largely at the experimental stage, and, in any event, are more in the nature of short-term palliatives than long-term control measures. They include sediment removal, aeration of bottom waters, and the addition of chemicals to precipitate phosphorus or oxidize the sediments.

The presence of large amounts of cyanobacterial or algal biomass is not invariably cause for concern, for these plants have many direct uses to man. Amongst the oldest is their use as food. The inhabitants of certain parts of Africa and of Mexico, for example, have long used *Spirulina*, a Cyanobacterium of moderately saline lakes, as a food base. Modern analysis has shown this plant to be particularly rich in proteins. Supplies are commer-

Fig. 9.8 Mass cultivation of algae being grown under pilot conditions at Bloemfontein, South Africa.

cially available in Australia under the trade name of 'Spirulina protein + '; all supplies, however, are imported directly from Mexico! Considerable attention is now being given to the culture of algae under controlled conditions (Fig. 9.8) as a food or food additive for man, fish or stock. In many situations, partially treated sewage is used as the base medium, though often the objective then is to increase the efficiency of sewage treatment rather than to produce algae.

Although there is some commercial production of algae for food, the mass cultivation of algae is mostly still at the pilot plant stage of investigation, and a number of important difficulties remain for resolution. Not the least of these are the small size of the principal algae involved (usually *Chlorella* and *Scenedesmus*, both green algae), and the dilute nature of the suspension. These difficulties provide problems for the efficient harvesting of algal crops. In sewage-derived media, the possible carry-over of pathogenic organisms is another difficulty.

Three further uses of algae are basically at the experimental stage. They are the use of algae to produce hydrocarbon fuels, as a source of glycerol and β-carotene, and as a biological surveillance and monitoring tool in water quality investigations. *Botryococcus braunii* is implicated in the production of the hydrocarbon fuel, for this widespread species synthesizes very large amounts of oil. Attempts to mass cultivate it and to understand better the nature of its field distribution and abundance are now in hand in Australia. *Dunaliella*

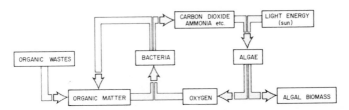

Fig. 9.9 Basic relationships between bacteria and algae in a waste stabilization pond.

salina is the alga being utilized in the production of glycerol and β-carotene. As noted previously, this alga synthesizes large quantities of glycerol as a means of surviving the high salinity of its environment (salt lakes). It also produces large quantities of β-carotene, a red pigment widely used in the food industry, and is claimed to be one of the richest plant sources of this compound. Diatoms are perhaps the most useful of those algae involved in water quality investigations.

A much more important use of Cyanobacteria and algae than any of those so far mentioned, and one well beyond the pilot or experimental stage, is their use in sewage purification. They play an important role in waste stabilization ponds or sewage lagoons, simple mechanisms for sewage treatment. There are many hundreds of such ponds throughout Australia. In them, organic waste material is degraded to simpler compounds so that the pond effluent is of much higher quality in terms of water management than the original sewage influent. Bacteria play the most important role in the degradation of the organic waste, but the Cyanobacteria and algae are needed to renew supplies of nutrients and oxygen (Fig. 9.9). Despite the proven value of waste stabilization ponds, surprisingly little is known about the ecological relationships involved and the biological functioning of the ponds.

10   Macrophytes

Surprisingly for so obvious a component of many aquatic communities, the large plants (macrophytes) of Australian inland waters have received relatively little attention from aquatic biologists. Indeed, it is less than a decade since the appearance of the first comprehensive flora or descriptive catalogue of them (Aston 1973: *Aquatic Plants of Australia*). There are many signs that this situation is rapidly changing; aquatic macrophytes are beginning to receive more attention from not only systematic botanists, but also ecologists interested in the role of macrophytes in the functional dynamics of Australian aquatic ecosystems. Impetus derives from many sources. Not the least is the economic significance of several species. Another is the greater recognition by limnologists of the ecological importance of macrophytes in many situations — an importance largely underestimated until recently. Especially in shallow water-bodies, they are important sources of energy, play a significant role in the cycling of nutrients and organic matter, provide food for herbivores, and contribute to detrital food chains. They also act as substrata for algae and invertebrates.

In this chapter, three topics are surveyed. The overall composition and distribution of the flora are discussed. Next, some salient ecological considerations are dealt with. Then the economic significance of macrophytes is considered.

COMPOSITION AND DISTRIBUTION

As a general rule, the macrophytes of inland waters are taken to include certain algae, namely the stoneworts (Charophyta), together with some mosses and liverworts (Bryophyta), some ferns and allied forms (Pteridophyta), and a variety of higher plants (Spermatophyta). Macrophytes are regarded as aquatic only if their principal foodmaking structures are permanently, or at least for substantial periods, submerged in, emergent from, or floating upon bodies of inland water. But the limits to what is and what is not an aquatic macrophyte are by no means precise. Many essentially terrestrial plants live near water and can survive inundation for a considerable time. Conversely, many essentially aquatic plants can survive out of water for extended intervals. Nevertheless, given a certain degree of abitrariness, aquatic plants as a whole do have a well-defined integrity as a part of many aquatic ecosystems.

The stoneworts (Charophyta) are large green algae represented by a single family, the Characeae, with half a dozen widespread genera, of which five occur in Australia: *Chara, Lamprothamnium, Nitella, Lychnothamnus* and *Tolypella* (sometimes partly amalgamated). All told, there are 29 species, mostly of *Chara* (Fig. 10.1(a)) and *Nitella*. Eight southern forms are endemic. Stoneworts occur throughout Australia, and occupy a high proportion of running and standing waters. They are completely submerged plants, often forming dense swards. In lakes, they may extend to depths well below those occupied by other aquatic macrophytes. Both fresh- and salt-water forms oc-

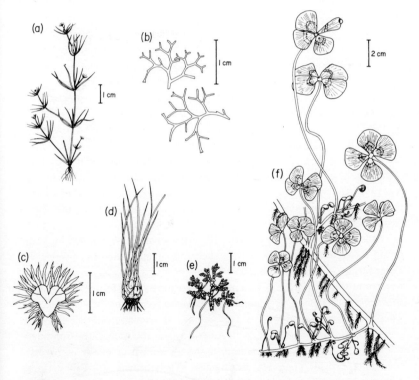

Fig. 10.1 Some aquatic charophytes, bryophytes and pteridophytes. (a) *Chara*; (b) *Riccia*; (c) *Ricciocarpus*; (d) *Isoetes*; (e) *Azolla*; (f) *Marsilea*. (a), (d)–(f) redrawn after Aston (1973), (b) and (c) after Cook *et al.* (1974).

cur, with *Lamprothamnium papulosum* known to survive well in a South Australian lake at a salinity twice that of sea water (and even higher salinities are tolerated by certain reproductive structures of this species). Stoneworts often have a characteristic smell, and are readily recognized because of this and their rather shiny translucent appearance, sometimes brittle nature, and the frequent occurrence of mucus. A few species, including some Australian ones, inhibit mosquito egg-laying or larvae.

Since mosses and liverworts (Bryophyta) are nearly always plants of damp places, the distinction between aquatic and terrestrial species is particularly hazy for this group of plants. Nevertheless, of the liverworts (Hepaticae), two are unequivocally aquatic, *Riccia 'fluitans'* and *Ricciocarpus natans*, neither being endemic to Australia. Riccia (Fig. 10.1(b)) is found either free-floating in the water column, floating at the surface, or attached to submerged surfaces. *Ricciocarpus* (Fig. 10.1(c)) floats at the surface, often together with *Azolla* (a water fern) or *Lemna* (a higher plant). Both liverwort genera characterize sheltered lowland standing waters rich in plant nutrients. Of mosses (Musci) frequently associated with free water in Australia, mention should be made of *Sphagnum* in bogs found in highland areas and *Sciaromium* in running waters.

Like liverworts and mosses, many ferns occur in damp places, and again it is sometimes difficult to decide whether a particular species is essentially aquatic or not (especially in wet tropical areas). There are many ferns, however, of which there is no doubt about their aquatic status. This is certainly true of *Isoetes* (Fig. 10.1(d)), a fern-ally (Lycopsida), which forms small grass-like tufts. Some of its species are totally submerged plants, though most are semi-aquatic or found in temporary waters. Note that horsetails (*Equisetum*), another well-known fern-ally with aquatic species, are not present in Australia. There is also no doubt about the aquatic status of many true fern species in Australia. Of these, four families in particular are important: Azollaceae, Marsileaceae, Parkeriaceae and Salviniaceae.

Azolla (Azollaceae; Figs 10.1(e), 10.2(a)) is a small (1–3 cm) floating plant found throughout Australia in rich, sheltered, standing bodies of water. Its lower surface is usually inhabited by a blue-green alga, *Anabaena azollae*, able to fix atmospheric nitrogen, with obvious benefits to both fern and alga. *Marsilea* (Fig. 10.1(f)) and, less important, *Pilularia*, represent the Marsileaceae. They are rooted forms with several aquatic or semi-aquatic species. Some species of *Marsilea* occur widely in Australia, and the sporocarps (reproductive structures) of *M. drummondii*, a species known generally as nardoo, sometimes provide food for Aborigines. When heavily grazed by stock, however, this species is toxic. *Ceratopteris*, the only aquatic genus of the Parkeriaceae, is free-floating, rooted in marginal mud, or rooted and submerged. *Salvinia* (Salviniaceae; Fig. 10.2(b)), on the other hand, is always quite free

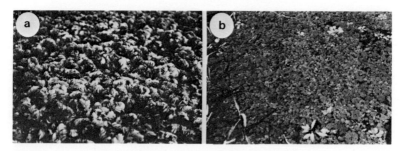

Fig. 10.2 Two floating ferns. (a) *Salvinia*; (b) *Azolla*.

of the bottom and floats at the surface of standing water-bodies. In these it may — and usually does — give rise to considerable problems because of excessive growth. An introduced plant, it is discussed in more detail later.

A much greater contribution to aquatic macrophyte communities in terms of species numbers and usually of biomass is provided by the flowering or higher plants (angiosperms). In total, there are aquatic representatives of 21 dicotyledonous families (29 genera, about 110 species) and a similar number of monocotyledonous families (45 genera, about 90 species) in Australia (Tables 10.1, 10.2). This contribution is far too large for higher plants to be discussed in the way the lower aquatic plants have been. Only a few of the more interesting genera can be mentioned: *Aldrovanda, Utricularia, Tristicha, Ruppia, Wolffia, Phragmites, Potamogeton* and *Nymphoides*. These have been selected for discussion because they illustrate the wide range of morphology shown by aquatic angiosperms and the diversity of environments inhabited.

Aldrovanda vesiculosa (Fig. 10.3(a)), the waterwheel plant or waterbug trap, is the only aquatic species of the Droseraceae. A northern, non-endemic species, it is distinguished by its carnivorous habits. Leaf blades are hinged, fringed with sensitive spines, and trap small aquatic animals by shutting very rapidly (0.02 seconds). The prey is subsequently digested. The waterwheel plant lies free-floating and suspended in the water column of still, shallow and generally nutrient-poor water-bodies.

Species of *Utricularia* (Fig. 10.3(b)), bladderwort, are also carnivorous. Their prey, however, is caught in small rounded bladders attached to much divided leaves. Like *Aldrovanda*, the trap is sensitive to touch; if a small animal stimulates the spines projecting from near the closed valve of the bladder there is a sudden inrush of water into the bladder. The prey is carried along by this inrush. *Utricularia* is widespread throughout Australia, with seven aquatic species recorded, some endemic. All are free-floating, rootless plants living in still, shallow water-bodies.

Table 10.1 Dicotyledonous macrophytes of Australian inland waters: composition, major lifeforms and gross distribution. Table derived and rearranged mainly from Aston (1973) and Specht (1981) but additional information from other sources is incorporated.

Family	Genus	Major lifeform*	Northern Australia	Southern Australia (including S.E. Qld)
Amaranthaceae	*Alternanthera*†	B.I.		+
Brassicaceae	*Nasturtium*†	B.I.		+
Calombaceae	*Brasenia*	B.II		+
Callitrichaceae	*Callitriche*†	B.I., B.III		+
Ceratophyllaceae	*Ceratophyllum*	A.II	+	+
Compositae	*Cotula*	B.I.		+
Convolvulaceae	*Ipomoea*	B.I.	+	
Crassulaceae	*Crassula*	B.I.		+
Droseraceae	*Aldrovanda*	A.II	+	
Elatinaceae	*Elatine*	B.III		+
Haloragaceae	*Haloragis*	B.I		+
	Myriophyllum†	B.I., B.III	+	+
Lentibulariaceae	*Utricularia*	A.II	+	+
Menyanthaceae	*Liparophyllum*	B.I		+
	Nymphoides	B.II	+	+
	Villarsia	B.I, B.II B.III		+
Nymphaeaceae	*Nelumbo*	B.I.	+	
	Nymphaea†	B.II	+	+
	Ondinea	B.I	+	
Onagraceae	*Ludwigia*	B.I	+	+
Podostemaceae	*Torrenticola*	B.III	+	
	Tristicha	B.III	+	
Polygonaceae	*Rumex*	B.I		+
Portulaceae	*Montia*	B.I		+
Ranunculaceae	*Ranunculus*	B.I, B.III		+
Scrophulariaceae	*Limnophila*	B.I	+	+
	Veronica†	B.I		+
Umbelliferae	*Hydrocotyle*	A.II		+
	Lilaeopsis	B.I, B.III		+

* A.I, free-floating at surface; A.II, free-floating below surface; B.I, rooted with part of plant emergent above water surface (helophytes); B.II, rooted with at least some leaves floating at surface; B.III, rooted with whole plant submerged. The notation follows Hutchinson (1975).
† Includes some naturalized species.

Table 10.2 Monocotyledonous macrophytes of Australian inland waters: composition, major lifeforms and gross distribution. Sources of information as in Table 10.1.

Family	Genus	Major lifeform*	Northern Australia	Southern Australia (including S.E. Qld)
Alismataceae	*Alisma†*	B.I		+
	Caldesia	B.II	+	+
	Damasonium	B.I		+
	Sagittaria†	B.I		+
Aponogetonaceae	*Aponogeton†*	B.I, B.II, B.III	+	+
Araceae	*Pistia*	A.I	+	+
Butomaceae	*Tenagocharis*	B.I	+	
Centrolepidaceae	*Centrolepis*	B.I, B.III	+	+
	Aphelia	B.I, B.III		+
Hydatellaceae	*Hydatella*	B.I, B.III		+
	Trithuria	B.I, B.III	+	+
Cyperaceae	*Eleocharis*	B.I	+	+
	'Scirpus' sens. lat.	B.I		+
	Cyperus	B.I	+	+
Eriocaulaceae	*Eriocaulon*	B.III	+	+
Hydrocharitaceae	*Blyxa*	B.III	+	
	Egeria†	B.III		+
	Elodea†	B.III		+
	Hydrilla	B.III	+	+
	Hydrocharis	A.I, B.II		+
	Maidenia	B.III	+	
	Ottelia	B.II, B.III	+	+
	Vallisneria	B.III	+	+
Juncaceae	*Juncus†*	B.I		+
Juncaginaceae	*Maundia*	B.I		+
	Triglochin	B.I, B.III	+	
Lemnaceae	*Lemna*	A.I, A.II	+	+
	Spirodela	A.I	+	+
	Wolffia	A.I		+
Lilaeaceae	*Lilaea†*	B.I		+
Najadaceae	*Najas*	B.III	+	+
Poaceae	*Arundo†*	B.I		+
	Phragmites	B.I	+	+
	Leersia	B.I	+	+
	Oryza	B.I	+	
Pontederiaceae	*Eichhornia†*	A.I	+	+
	Monochoria	B.I	+	
	Pontederia†	B.I		+
Potamogetonaceae	*Potamogeton*	B.II, B.III	+	+

Table 10.2   (Contd.)

Family	Genus	Major lifeform*	Northern Australia	Southern Australia (including S.E. Qld)
Restionaceae	*Leptocarpus*	B.I	+	
Ruppiaceae	*Ruppia*	B.III		+
Sparganiaceae	*Sparganium*	B.I		+
Typhaceae	*Typha*†	B.I	+	+
Zannichelliaceae	*Lepilaena*	B.III		+
	Zannichellia	B.III		+

* For explanation of lifeform notation see footnote to Table 10.1.
† Includes some naturalized species.

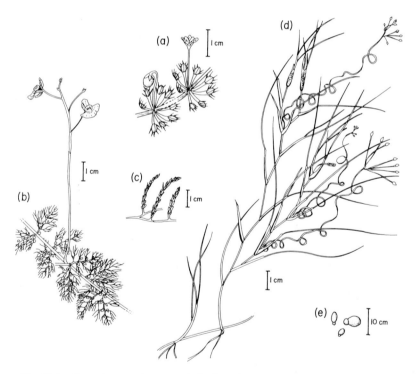

Fig. 10.3   Some interesting aquatic higher plants (angiosperms). (a) *Aldrovanda vesiculosa*; (b) *Utricularia*; (c) *Tristicha*; (d) *Ruppia*; (e) *Wolffia*. (a), (b), (d) and (e) redrawn after Aston (1973), (c) after Cook *et al.* (1974).

Tristicha (Fig. 10.3(c)), with a single, non-endemic species, *T. trifaria*, together with *Torrenticola queenslandica*, represents the family Podostemaceae in Australia. Members of this family are unusual in that they look like mosses, liverworts, algae or lichen. *Tristicha* itself forms a dense, submerged, moss-like sward on rocks. Like all members of the family, the species is confined to flowing waters, including swiftly flowing streams; its growth habit is an obvious adaptation to life in such environments. So far the plant has been found only in the Northern Territory and Western Australia.

Ruppia (Fig. 10.3(d)) has quite a different appearance and habitat. Its species are grass-like and usually occur in slightly to markedly saline lakes. In these, some species have been recorded at salinities of 230°/oo (a salinity over six times that of seawater), so that *Ruppia* has the greatest tolerance to salinity of any aquatic macrophyte. The internal accumulation of certain organic substances, e.g. the amino acid proline, provides the means whereby such high salinities can be tolerated. Other adaptations to the harsh and fluctuating aquatic environment of salt lakes include adapted lifecycles and reproductive patterns. Thus, the four Australian species include both perennial and annual species, and reproduction may be by asexual turions (dwarf dormant shoots) and/or sexually produced seeds.

Species of *Wolffia* (Lemnaceae; Fig. 10.3(e)), tiny duckweeds, are the smallest of all flowering plants, with individuals often scarcely visible (<2 mm), yet sometimes sufficiently numerous to form a complete cover to the water surface of their habitat. Widespread throughout Australia, two species occur here, one of which is endemic. Both are tiny, rootless plants which float on the surface of sheltered, lowland, standing bodies of water such as farm dams and billabongs, frequently with other, larger duckweeds (*Lemna, Spirodela*).

At the opposite end of the size scale to *Wolffia* lies *Phragmites* (Poaceae; Fig. 10.4), the common reed. Its commonest Australian species, *P. australis*, is a robust emergent plant that can reach a height of over 3 m. An introduced relative, *Arundo donax*, the giant reed, can even attain a height of 6 m! There are two species of *Phragmites* in Australia, neither of which is endemic. Found throughout the continent, including inland arid regions, *Phragmites* often occurs in dense stands fringing the edges of creeks, rivers, drains, swamps and lakes wherever there is sufficient shelter, mud, and lack of current and water level change to support it. *Phragmites australis* is the most common species in the southern half of the continent; *P. karka* is the northern species.

The genus *Potamogeton* (Potamogetonaceae; Fig. 10.5(a)), with several species, is widespread throughout Australia, and has been recorded from fresh to slightly saline, and still to strongly flowing inland waters. Reference is made to it here because it illustrates well a major feature of many aquatic

Fig. 10.4 A large stand of *Phragmites australis* fringing a permanent waterhole near Lake Eyre South, South Australia.

macrophytes, namely, the occurrence of variation in morphology within an *individual* plant according to which part of the plant is submerged and which emergent. Figure 10.5(a) illustrates this feature in *P. tricarinatus*.

Nymphoides (Menyanthaceae; Fig. 10.5(b)), like *Potamogeton*, has several species in Australia and is recorded from all states. It is, however, more common across the northern part of the continent. It is mentioned here because its species provide examples of an important lifeform of aquatic macrophytes that is not illustrated by any of the previous examples considered; plants are rooted in the bottom sediments, but have leaves which float at or just below the water surface.

Finally, in any consideration of angiosperms it should be emphasized that numerous species, whilst essentially terrestrial plants, are nevertheless characteristically associated with inland waters and wetland environments. No adequate discussion of these is offered here, but any book concerned with Australian inland waters would be incomplete were it not to make at least passing mention of the river red gum (*Eucalyptus camaldulensis*) and paperbarks (*Melaleuca*). The river red gum, a large, majestic tree, is found bordering almost all lowland watercourses in Australia, even in the arid zone; it is absent only in the southwest and the Nullarbor Plain. The dominant tree nearest the water, it may also extend for considerable distances over floodplains. Flooding appears to be a necessary prerequisite for the germination and growth of sap-

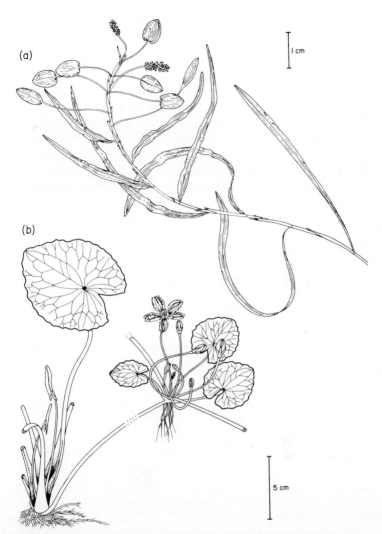

Fig. 10.5 Two further aquatic higher plants (angiosperms) of interest. (a) *Potamogeton tricarinatus*; (b) *Nymphoides crenata*. Redrawn after Aston (1973).

lings. Unfortunately, introduced willow trees (*Salix*) have replaced it over long distances near some rivers, e.g. the Murray River. Paperbarks include species of a size ranging from bushes to small trees. They fringe watercourses in many parts of Australia, but are particularly characteristic and conspicuous near watercourses and swamps in northern Australia.

Turning now to a consideration of the distribution of aquatic macrophytes, it should first be noted that many aquatic plants have extremely wide distributions, with several being cosmopolitan or ubiquitous. Quite a few cosmopolitan species occur naturally in Australia, with the common reed (*Phragmites australis*) perhaps the most notable example. Some others are the common duckweed (*Lemna minor*), hornwort (*Ceratophyllum demersum*) and fennel pondweed (*Potamogeton pectinatus*). Most species of Australian aquatic macrophytes, however, have a more restricted distribution; some 55% of non-angiosperm species are endemic, as are some 60% of dicotyledonous angiosperms and some 30% of the monocotyledons. Overall, one in two of all native Australian aquatic macrophyte species is endemic. Most genera, on the other hand, are widespread, with only four confined to Australia, namely *Ondinea, Maidenia, Maundia* and *Damasonium*, and a further one, *Lepilaena*, found outside only in New Zealand. This is in marked contrast to the position in terrestrial plants.

Within the continent, many genera and species, both endemic and otherwise, are widespread; many others are regionally restricted or of rather local distribution. As well, there is often considerable variation between similar environments in the same general area in terms of dominant species, community composition and seasonal succession. Whatever the case, analysis of biogeographical relationships indicates that aquatic macrophytes in northern Australia have strong affinities with the African-Indo-Malayan flora and the flora of other tropical regions, whereas those in the south have their closest affinities with the flora of other southern continents. The distribution of aquatic angiosperm genera in northern and southern Australia is also indicated in Tables 10.1 and 10.2

Worldwide patterns of plant distribution and the analysis of biogeographical relationships are considerably hampered by the presence in all countries of naturalized (adventive) species. Those aquatic species known to have been naturalized in Australia since 1788 are listed in Table 10.3. A few species which are only doubtfully naturalized are not listed, nor are the many introductions into home ponds and aquaria that do not have self-sustaining wild populations. Also not listed are numerous exotic species which, whilst not aquatic in the sense of that word used here, are nevertheless often found associated with inland waters as bankside vegetation. The occurrence of *Salix* has already been mentioned. Another introduced bankside species is the blackberry (*Rubus*). In many parts of temperate Australia this plant is spreading rapidly, particularly along watercourses, and proving to be a considerable nuisance and a noxious species.

It is likely that many more adventive aquatic macrophytes will appear in the next 200 years. Note that some of them have proved to be considerable

Table 10.3 Aquatic macrophytes naturalized in Australia. Table based upon Aston (1973) and update.

Family	Species
Ferns	
Salviniaceae	*Salvinia molesta* (salvinia)
Dicotyledons	
Amaranthaceae	*Alternanthera philoxeroides* (alligator weed)
Brassicaceae	*Nasturtium microphyllum* (one-row watercress)
	N. officinale (watercress)
Callitrichaceae	*Callitriche hamulata* (starwort)
	C. stagnalis (common starwort)
Haloragaceae	*Myriophyllum aquaticum* (parrot's feather)
Nymphaeaceae	*Nymphaea capensis* (African waterlily)
	N. flava (waterlily)
Onagraceae	*Ludwigia palustris* (marsh Ludwigia)
Scrophulariaceae	*Veronica anagallis-aquatica* (blue water speedwell)
	V. catenata (pink water speedwell)
Monocotyledons	
Alismataceae	*Alisma lanceolatum* (water plantain)
	Sagittaria engelmanniana
	S. graminea (sagittaria)
	S. montevidensis
	S. sagittifolia (arrowhead)
Aponogetonaceae	*Aponogeton distachyon* (Cape pond lily)
Hydrocharitaceae	*Egeria densa* (dense waterweed)
	Elodea canadensis (Canadian pondweed)
Lilaeaceae	*Lilaea scilloides* (Lilaea)
Pontederiaceae	*Eichhornia crassipes* (water hyacinth)
	Pontederia cordata
Typhaceae	*Typha latifolia* (bulrush)

nuisances; *Eichhornia crassipes* (water hyacinth), *Salvinia molesta* (salvinia) and *Alternanthera philoxeroides* (alligator weed), in particular, have cost Australia many millions of dollars in efforts to control excessive growth in reservoirs, rivers and channels. This is discussed further below.

The composition of the aquatic flora is also likely to change in another direction—some species are likely to become extinct as a result of man's impact upon their habitats. This subject is discussed in chapter 12. Already, some 18 Australian species of aquatic macrophyte must be regarded as threatened in this way (see Table 11.1), including some of the species of angiosperm discussed above as being species of particular interest (*Aldrovanda, Tristicha*). More by good fortune than by good management, no species so far as known has yet become extinct.

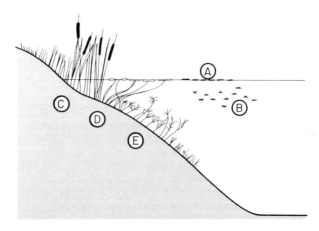

Fig. 10.6 Major lifeforms of aquatic macrophytes. A, free-floating at surface; B, free-floating beneath surface; C, emergent; D, with floating leaves; and E, submerged.

ECOLOGICAL CONSIDERATIONS

The structure and form of aquatic macrophytes fall rather easily into five main categories or 'lifeforms' (Fig. 10.6). Two of these include floating plants with roots suspended in the water column, and three, plants rooted in bottom sediments. Of floating forms, there are those that float at the water surface (Fig. 10.6A), and those that float suspended in the water column itself (B). Of rooted forms, some have parts which emerge from the water (C), others have leaves that float at the water surface (D), and others are fully submerged (E). All five lifeforms are well-represented in Australia — though apparently not in nearby New Zealand which seems to have a dearth of fully submerged native forms. Each lifeform (and often different species of the same lifeform) has associated differences in the invertebrate community inhabiting it. Tables 10.1 and 10.2 indicate the major lifeform of all Australian angiosperm genera.

The major lifeforms often display a zonation in permanent standing fresh waters of the general pattern illustrated in Figs 10.6 and 10.7. There is a zone near land of emergent species, one a little deeper of floating-leaved species, and a deep zone of fully submerged or free-floating species. The lowest depth to which macrophytes descend is determined largely by light penetration in deep water-bodies (Fig. 10.8). Many Australian lakes or other standing water-bodies display this zonational pattern or some variant of it (Fig. 10.9). However, Australia is characterized amongst other climatic phenomena by its high rainfall variability, with resultant unstable levels in many water-bodies.

Fig. 10.7 Lake Euramoo, Queensland. The zonation of emergent and floating macrophytes is clearly visible as marginal bands of vegetation.

Consequently, clear zonations of aquatic macrophytes are often *not* discernible, and, in fact, many water-bodies with unstable or astatic levels lack any well-defined macrophyte community altogether. The latter is also often the

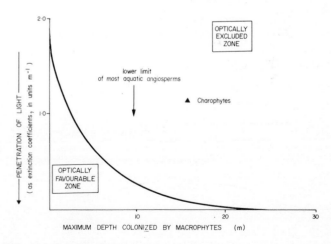

Fig. 10.8 Relationship between light penetration (represented as extinction coefficients) and the maximum depth of aquatic macrophyte colonization. Redrawn after Ganf (1977).

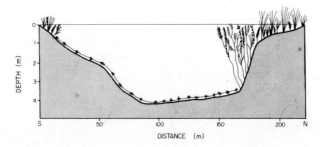

Fig. 10.9 Vegetation profile diagram of Fresh Dip Lake (a freshwater lake), South Australia, along north–south transect. Redrawn after Brock (1979). Reeds: Gahnia, *Lepidosperma* spp., *Machaerina*, '*Scirpus*' sens. lat.; *Myriophyllum propinquum*; *Potamogeton pectinatus*; *Chara vulgaris*; *Ruppia*.

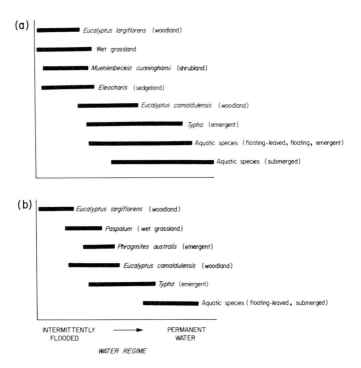

Fig. 10.10 Generalized zonation of plant communities according to water regime in two wetland environments in southeastern Australia. After various authors but based on Briggs (1981). (a) Swamps, northern Victoria; (b) Macquarie Marshes, NSW.

case in storage reservoirs; here drawdowns to release water for a variety of purposes are usually inimical to the development of at least rooted aquatic macrophytes.

Zonation is also usually indistinct in shallow water-bodies such as marshes, swamps and other wetland habitats. In these it is the periodicity and availability of water that largely determine the composition and structure of plant communities, not water depth. Figure 10.10 illustrates this point. Note the extensive zones of overlap between the various communities. Likewise, macrophyte zonation is not an obvious phenomenon in salt lakes. Submerged macrophytes in these environments, if present (and often they are not), are almost exclusively charophytes (*Lamprothamnion*) and species of *Ruppia* or *Lepilaena*. In shallow salt lakes they are generally interspersed; in deeper salt lakes, zonation — at best — is rather weak (Fig. 10.11).

In rivers and streams, macrophyte distribution is ordered by a quite different set of factors from those operating in standing waters. In running waters, it is usually not depth nor the availability of water, but current speed, that is the main determinant of macrophyte occurrence and abundance. The effect is principally exerted by the determination of substratum type; slowly flowing waters are depositional environments and produce silty substrata, whereas rapidly flowing waters are erosional environments and have rocky substrata. Since few macrophytes can gain nourishment from rocks and settle and grow in strong currents, there are few of them in rapidly flowing streams. In southern Australia, those that do occur in such environments are mosses or liverworts, but in a few isolated streams in northern Australia two species of the Podostemaceae have also been recorded (*Torrenticola queenslandica, Tristicha trifaria*). With decreasing current velocity, streams and rivers provide increasing opportunities for the development of aquatic macrophytes, until in

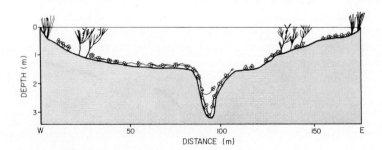

Fig. 10.11 Vegetation profile diagram of Little Dip Lake (a saline lake), South Australia, along east-west transect. Redrawn after Brock (1979). Ⲽ *Juncus kraussii, Gahnia filum;* ⟞⟞ *Lamprothamnium;* Ⲽ *Ruppia.*

the slowest currents the opportunities are at least as great as those provided by standing waters.

As a general rule, zonation of river macrophytes is not a notable feature; plant communities in rivers and streams are typically a rather unstable and complex mosaic of species ranging along the length of the river and in time. It is only in the quietest reaches and in billabongs that an approach to lake zonational patterns occurs. Again, however, the variability in river levels throughout most of Australia militates against extensive macrophyte development. Very little of the edges of the River Murray, for example, possesses extensive beds of vegetation, despite the relatively slow speed of water movement. Some of the more common macrophytes of Australian rivers and streams include species of *Chara, Nitella, Ceratophyllum, Myriophyllum, Ranunculus, Aponogeton* and *Potamogeton*, as well as *Vallisneria spiralis* and *Triglochin procera*. They are characterized by their strong holdfast organs (roots in non-algal species) and a pliant robust form capable of withstanding turbulent, unidirectional water movement. All species of angiosperms involved, it is added, may also be found in still waters, if somewhat infrequently in some cases; indeed the only angiosperm species confined to running waters are those in the Podostemaceae.

Finally, in this consideration of ecological matters, it must be reported that at present little can be said about energetic and nutrient relationships based upon work on Australian aquatic macrophyte communities. Studies of the rates of production, decomposition and nutrient cycling are just commencing here. Since, however, productivities of Australian macrophytes probably do not differ greatly from macrophytes elsewhere, it is of interest to compare the productivities of various sorts of plant community (Fig. 10.12). As can be seen, emergent macrophyte communities are by far the most productive. In many parts of Australia subject to flood–drought regimes of water availability, especially large productivities are no doubt a feature of freshly flooded areas where nutrients are plentiful following release during the accelerated decomposition of macrophytes during the previous dry period. In any event, of the four major determinants of plant productivity — nutrient supply, water, temperature and sunlight — two at least are not likely to be limiting over large areas of the continent.

ECONOMIC SIGNIFICANCE

Aquatic macrophytes have considerable economic significance. On the one hand, they may give rise to a number of major problems; on the other, they have a number of actual or potential uses.

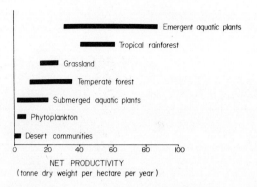

NET PRODUCTIVITY
(tonne dry weight per hectare per year)

Fig. 10.12 Productivities of various plant communities compared. Rearranged after various authors.

The problems caused by aquatic macrophytes in Australia vary in intensity according to locality, season and species, but in general they are all a reflection of excessive and unwanted plant growth. Some of these problems are: decreased flows in rivers, drainage ditches, irrigation channels and other water supply conduits, leading to flooding or diminution of supplies of water; impeded boating and swimming; reduced recreational values; lower water storage capacities; increased siltation rates; inoperable engineering structures such as floodgates and locks; and more difficult access to water-bodies. There are a great many more, some of which, though minor in general, may be very important locally.

Many of these problems are most intense and obvious in man-made watercourses or impoundments. For example, the impact of decreased flows is particularly important in the irrigation areas of southeastern Australia, and water storages in Queensland are plagued by floating plants (Fig. 10.13). It is difficult, if not impossible, to put a precise figure on the cost to the nation of water weed problems, but the annual cost of control measures—clearly an underestimate of total cost—approaches $3–5 million. Undoubtedly control costs will rise in the near future due to the inflating cost of herbicides, labour and other elements of control.

Some 30-odd species have been identified as water weeds in Australia. Not all, of course, are of equal importance, but those that *overall* create the greatest nuisance are Canadian pondweed (*Elodea canadensis*), milfoil (*Myriophyllum*), pondweeds (*Potamogeton*), water hyacinth (*Eichhornia crassipes*), salvinia (*Salvinia molesta*), alligator weed (*Alternanthera philoxeroides*), common reed (*Phragmites australis*) and cumbungi (*Typha*). Several plants are also important as major bankside weeds; water couch (*Paspalum*)

F

Fig. 10.13 Excessive growth of water hyacinth choking lagoon near Mount Garnett, northeast Queensland, February 1965.

and blackberry (*Rubus*) are perhaps the most important of these. Water couch is a native, perennial grass that grows out from banks onto the water surface. Both native and naturalized species are involved, but it will be noticed (Table 10.3) that several of the major pest species are naturalized forms. With care and more rigorous quarantine laws, therefore, several of Australia's weed problems *could* have been averted.

Quite why aquatic macrophytes create problems has many answers. Not all can be considered here. In some cases problems arise because man has added to inland waters large amounts of sewage, and this contains plant nutrients (see chapter 12 for a discussion of eutrophication). The excessive growth of ribbon-weed, curly pondweed and Canadian pondweed in Lake Burley Griffin, Canberra, provides an example. In other cases, problems arise because of man's alteration of hydrological regimes and his creation of new water-bodies; these events often result in especially favourable conditions and environments for macrophyte growth. Examples of problems that have arisen in this way are given by nuisance growths of the common reed, cumbungi, water couch and Canadian pondweed in irrigation areas of southeastern Australia. Similarly, curly pondweed, water thyme (hydrilla), salvinia and the water hyacinth are particularly troublesome weeds in many impoundments of the warmer parts of eastern Australia. In still other cases, it appears that naturalized weeds grow more vigorously than native species, or in a few situations have invaded an ecological niche otherwise not fully utilized by native species. Man exacerbates this problem by being the main dispersal agent.

Several measures can be used to control water weed problems. The oldest is that of manual or mechanical control; weed growth is simply cut or otherwise directly removed. A more recent measure is control by herbicides. Here, synthetic organic poisons are used to limit growth or kill the weed involved. Whilst often efficacious, there are some important potential dangers in the use of herbicides, including the possibility of damage to non-target aquatic life and the environmental persistence of the herbicide (with consequent dangers associated with the later use of water). Biological control is another and more desirable method of control, but is presently of rather limited applicability in Australia. However, spectacular control of salvinia on Lake Moondarra (near Mount Isa, Queensland) has been achieved by use of the introduced weevil, *Cyrtobagous singularis* (though other environmental factors probably helped). And some control of water hyacinth is being achieved by an imported South American weevil (*Neochetina eichhorniae*), and of alligator weed by another imported beetle (*Agascicles hygrophila*). Other insect herbivores are being considered for introduction into Australia as possible control measures. The herbivorous grass carp (*Ctenopharyngodon idella*) has also been considered in this way (but see comments in chapter 5). Manipulation of certain environmental factors, namely light, substrate and water level, represents yet another control measure of some value, especially when used in conjunction with other measures. Finally, mention should be made of one basic control measure: prevention. Many problems can be avoided or alleviated by early steps against the build-up of nuisance growth or, better, by preventing colonization.

While excess macrophyte growth creates problems in some situations, in others it is a desirable event, for there are many uses to which macrophytes can be put. It must be added, however, that the extent to which these are exploited in Australia remains presently insignificant. The main direct uses are as a food for man and livestock, for the extraction of protein, as a fibre, for the manufacture of yeasts, alcohol and other useful by-products, to make mulches, composts and fertilizers, and to generate methane. As a food for man, it may be recalled that rice (*Oryza sativa*) is the world's most important crop plant and is usually cultivated as an aquatic annual plant. It is grown under irrigation in many areas of Australia. As livestock food, macrophytes are less useful since a major difficulty is the removal of excess water without denaturing the plant. Further research, it is expected, will overcome this. As a source of fibre, some species of macrophyte have long been useful, and in some European countries, especially Rumania, vast areas of reedswamp are exploited commercially to produce cellulose fibre — as well as alcohol, yeasts and other products.

An important indirect use of macrophytes that is attracting increased attention is their use as biological mechanisms for the removal of plant nutrients from wastewaters. Emergent macrophytes are used for this purpose at several locations in Europe and have proved to be most useful. Investigations are now in hand on a number of other continents, including Australia (Fig. 10.14), to determine the best operational techniques under local conditions. Early evidence clearly indicates that, with proper management, aquatic macrophytes can provide an extremely simple, cheap and efficient way of treating wastewater to decrease plant nutrient content. In this regard, it may also be noted that many aquatic macrophyte species aid water clarification, and are tolerant to, and can remove a number of, water pollutants other than plant nutrients, e.g. phenols, insecticides, and some heavy metals such as mercury and cadmium. A few macrophytes are known that even produce antibacterial exudates and that can therefore directly lower bacterial counts in wastewater. The ease of harvesting macrophytes may be of some significance in their use as a wastewater purification technique, though the ease of harvesting is probably less important than the capacity of the whole system to act as a sink.

Fig. 10.14 *Typha* growing in experimental ponds, CSIRO Division of Irrigation Research, Griffith. The investigation is concerned with the use of *Typha* in wastewater purification.

11   Conservation

Man's impact on inland waters and their biota has been profound: few water-bodies, worldwide, are likely to remain completely unaltered by the end of this century. The impact is as old as civilization, but it is the past century that has witnessed the greatest changes, and only the past few decades that have seen the development of an awareness of conservation matters, and a realization that these changes have important implications for man.

This chapter considers conservation within the context of Australian inland waters. The intention is to provide not an overview of conservation, but rather a summary of particular regional features and needs. It does not deal with the nature of man's impact (see chapter 12) except when special cases are involved. Some recent books dealing with general matters are listed in the references.

THE MEANING OF CONSERVATION

Conservation has come to mean different things to different people and it is as well to begin by providing a definition. A recent authoritative one (by the International Union for the Conservation of Nature and Natural Resources (IUCN), 1980) is:

> 'The management of human use of the biosphere so that it may yield the greatest sustainable benefit to present generations while maintaining its potential to meet the needs and aspirations of future generations.'

This definition suffices as a working one for present circumstances, though it may be noted that some conservationists have reservations about it on the basis that it effectively perpetuates the notion that man is apart from the natural world (a dangerous idea), rather than part of it. It is important to note that the IUCN definition has man as the direct beneficiary of conservation, and that conservation is concerned with both present and future generations of mankind. Implicit in the definition is the need to prevent species extinction and maintain ecosystem viability. Thus, fundamentally, *conservation is concerned with the intelligent management of natural resources for man's benefit.*

Conservation, as defined here, is not synonymous with the term as sometimes used by many Australian government agencies (both federal and state) and engineers who use it in the sense of 'resource husbandry'. Nor does it correspond to the layman's frequent view of it: an emotional response to the imprecations of biologists to preserve rare animals and plants and pleasant natural areas, or worse, an environmental involvement by the socially affluent and elite. It is important, therefore, to provide an explicit statement about the actual benefits to mankind of aquatic animals, plants and ecosystems. These benefits constitute the pragmatic rationale for conservation.

Seven clear benefits or values can be recognized: economic, recreational, aesthetic, cultural, scientific, educational and ecological.

The *economic value* of inland waters is immense. To a significant degree it is greatest when the waters involved are as natural as possible. Fisheries, water supplies for domestic, stock and industrial use, and irrigation water are some of the more obvious economic ends served by lakes and rivers. Lakes and rivers also have considerable *recreational value*, a value that has become more important in recent decades. Recreational fishing, boating, water skiing, swimming and duck shooting are all water-based activities of increasing popularity with contemporary Australians. Not all such activities, of course, require *natural* aquatic environments for their pursuit, but initial pressures almost always bear upon the natural environment. When the demand cannot be met by natural water-bodies, man-made lakes may be used, and there are already schemes afoot to construct near Melbourne and Sydney large man-made lakes to satisfy local demands for water-based recreation.

Related to the recreational benefits offered by natural waters is their *aesthetic value* — sometimes termed 'inspirational' or 'spiritual' values. It is difficult if not impossible to quantify these, but lakes, rivers and streams have long attracted the attention of artists, poets, writers and musicians — firm evidence of their evocative effect. Indeed, it has been said that nature conservation penetrates the emotional and subconscious; it touches the roots of human nature and is ultimately to do with being and feeling (Ratcliffe 1976). These same feelings have led to the construction of artificial lakes specifically for aesthetic reasons; Lake Burley Griffin, for example, was built to provide an aesthetic focus for the urban environment of Canberra. Lake Burley Griffin is undoubtedly a pleasant feature of the Australian Capital Territory, but natural water-bodies surely have higher aesthetic values than those created by man; who would claim that the new Serpentine-Huon impoundment in southwestern Tasmania is aesthetically more pleasing than the lake it destroyed, the incomparably beautiful Lake Pedder (Fig. 11.1), or that the proposed impoundment of the Franklin River in the same region will match the wild grandeur of the untouched river and its valley? Also not to be forgotten here is the aesthetic appeal of many aquatic animals and plants: graceful

Fig. 11.1   Lake Pedder before its destruction. Of this lake, *Project Aqua* said: 'Its impending destruction to provide power production for about half a century must be regarded as the greatest ecological tragedy since European settlement of Tasmania.' It was flooded in the early 1970s by the Hydro-Electric Commission of Tasmania.

waterbirds, placid reedswamps, and many aquatic insects. The *cultural* importance of inland waters is another unquantifiable but significant benefit: lakes and rivers have played critical roles in the history of mankind and form an integral part of our cultural heritage.

The *scientific* and *educational values* of inland waters and their biota are themselves manifold. Scientifically, the aquatic biota offers a wide diversity of forms of interest to a variety of disciplines: ecology, biogeography, physiology, genetics, and many others. It also offers a rich source of genetic material for future study. As natural ecosystems, the biota and its environments provide natural baselines against which to measure pollution elsewhere, from which to derive management guidelines for disturbed environments, and in which to study ecosystem processes in undisturbed environments. Educationally, natural water-bodies are important because they offer discrete but cohesive parts of the biosphere for observation by the increasing number of people interested in biology and other sciences.

Finally, the *ecological value* of inland waters should be stressed. They are an integral part of the life-support systems of the planet, systems without which man could not survive. While perhaps the most difficult to appreciate of all the values put forward as part of the conservation rationale, in the long haul this value may prove to have been by far the most significant.

THE CONSERVATION STATUS OF THE AQUATIC BIOTA AND ENVIRONMENTS

Notwithstanding the many benefits, a great deal remains to be done to ensure the adequate conservation of Australian aquatic environments and their biota. Many aquatic plants and animals are in danger of extinction. Several types of aquatic ecosystem are not adequately protected within national parks or similar areas (in some states, national park authorities do not even have control of aquatic ecosystems in parks!). And damage and threats of damage to the aquatic biota and habitats continue apace. This state of affairs persists in spite of an increased interest by the scientific community in the Australian aquatic biota, a concern for its possible destruction, an appreciation of its uniqueness, and recognition of Australia's international conservation obligations. It may help to focus attention on this distressing situation if the conservation status of the Australian aquatic biota and aquatic environments is surveyed in the light of current knowledge.

The status of the biota

The terms 'endangered', 'vulnerable' and 'rare' are used here in general accord with the definitions of IUCN (1980): an 'endangered' species is one in danger of extinction and whose survival is unlikely if threats to its existence continue; a 'vulnerable' species is one not yet endangered but likely to be if threats continue; and a 'rare' species is one whose world population is small and at risk, but which is not yet endangered or vulnerable. Whilst the process of species extinction is, of course, a normal evolutionary phenomenon under natural circumstances, man has greatly accelerated its rate and altered its scope; a recent estimate for the present rate for animals alone is that it is 1000 times greater than in the period immediately before man. The numbers of endangered species indicate that we stand to lose substantially more of our natural heritage within the next few decades.

Most species of algae are cosmopolitan, and so few if any are endangered, vulnerable or rare. There are some endemic species in Australia, but so far as known none is at risk. A different situation prevails for aquatic macrophytes. The conservation status of all Australian rare or threatened species of pteridophyte, gymnosperm and angiosperm has recently been listed (Leigh *et al.* 1981), and in the list are 18 aquatic species. These are shown separately in Table 11.1. Most of the plants in the table are endemic to Australia, but some

Table 11.1 Rare or threatened species of aquatic macrophytes in Australia. Derived from Leigh *et al.* (1981).

Family/species	Distribution	Occurrence*	Conservation status
Nymphaeaceae			
Nelumbo nucifera	NT, Qld	3†	Vulnerable
Ondinea purpurea	WA	3	Rare
Droseraceae			
Aldrovanda vesiculosa	NT, Qld, NSW	3†	Vulnerable
Podostemaceae			
Torrenticola queenslandica	Qld	2†	Rare
Tristicha trifaria	NT	2†	Vulnerable
Apiaceae			
Hydrocotyle lemnoides	WA	2	Vulnerable
Menyanthaceae			
Nymphoides exigua	Tas.	2	Vulnerable
N. furculifolia	NT	2	Uncertain
N. stygia	SA	1	Extinct
Villarsia calthifolia	WA	2	Rare
V. congestiflora	WA	2	Uncertain
V. lasiosperma	WA	3	Endangered
V. submersa	WA	2	Endangered
Callitrichaceae			
Callitriche brachycarpa	Vic., Tas.	3	Uncertain
Hydrocharitaceae			
Vallisneria caulescens	Qld	3	Vulnerable
Aponogetonaceae			
Aponogeton hexatepalus	WA	2	Vulnerable
Zosteraceae			
Zostera mucronata	WA, SA	3	Uncertain
Zannichelliaceae			
Zannichellia palustris	SA, NT	2†	Rare

* 1, known from one locality; 2, restricted occurrence (< 100 km range); 3, small populations in specific habitats (> 100 km range).
† Occurs outside Australia.

are of interest on grounds other than endemicity. Included, for example, is *Aldrovanda vesiculosa*, a carnivorous plant and the only aquatic species of the Droseraceae (sundews) in Australia. Also included are the only two Australian representatives of the family Podostemaceae, *Torrenticola queenslandica* and *Tristicha trifaria*; both occur in northern Australia and are stongly adapted for life in running waters, with a body that is lichen-, moss- or algal-like in general form.

Knowledge of the conservation status of Australian aquatic invertebrates is less well-developed, although already some species are known definitely to be in danger of extinction and a few to have become extinct. Since many insect and crustacean groups in particular have numerous endemic species of localiz-ed distribution in areas presently subject to man-made change, it may safely be presumed that the actual status is much worse than we realize. Uncertainty about their conservation status will undoubtedly continue for many years in the light of our continued ignorance about the distribution and abundance of most Australian aquatic invertebrates. In this respect, Australia is little different from other countries; it is only very recently, for example, that preparation could begin of lists of invertebrates for inclusion in official records of endangered species (i.e. the so-called Red Data books prepared by the IUCN).

Amongst groups of aquatic invertebrates in Australia known to have elements at risk, special mention may be made of the Mollusca, Plecoptera, Odonata, Trichoptera, Blephariceridae, Syncarida and Parastacidae. Other groups containing elements at risk or with a strong possibility of doing so include the Tricladida, Hemiptera, Cladocera, Ostracoda, Branchiura and Phreatoicidea.

Of the molluscs, both the Bivalvia and Gastropoda are known to have threatened species. Thus, *Hyridella glenelgensis*, a freshwater mussel confined to the Glenelg River system in western Victoria, is very difficult to find and must be regarded now as a vulnerable species. *Vivipara sublineata*, a freshwater snail of the Murray–Darling system, should be regarded as en-dangered; it appears to be extinct in the Murray itself. *Glacidorbis pedderi*, an unusual hydrobiid snail, is certainly an endangered species. Known previously from only Lake Pedder and the lower Gordon River area of Tasmania, it did not survive the inundation of Lake Pedder, and now the lower Gordon River also faces inundation.

In several presently recognized species of the Plecoptera, isolated popula-tions are frequently different from central ones, with differences that may prove to be of subspecific (or even specific) rank. At present, several of these isolated populations are at risk because of dams (e.g. *Trinotoperla 'nivata'* in the Grampians, Victoria), restricted distributions (e.g. *Leptoperla 'neboissi'* in

South Australia, known from only one stream), or agriculture (e.g. *Riekoperla* '*rugosa*' also in South Australia and confined to a few isolated wooded knolls on the Eyre Peninsula). There are also several species whose entire populations are confined to small areas, and because of this are best regarded as rare species, though perhaps not immediately at risk. Such are *Riekoperla darlingtoni* confined to the summit of Mt Donna Buang, Victoria, and *Leptoperla cacuminis* confined to the summit of Mt Kosciusko, NSW. All species of the genus *Thaumatoperla* also have restricted distributions.

Of odonate species threatened, particular note is made of *Hemiphlebia mirabilis*, the only extant species of the ancient family Hemiphlebiidae. This species is confined to a few localities in Victoria, where until recently it was thought to have died out. It was rediscovered a little while ago but its existence continues to be threatened by urbanization, and it is clearly an endangered species. Several other species of dragonfly confined to small localized areas also appear to be under threat. Vulnerable species include *Nososticta pilbara* and *Ictinogomphus dobsoni* known only from the Millstream Springs on the Fortescue River, northwestern Western Australia. The underlying aquifer is being pumped to provide industrial supplies, with obvious risks for surface water levels. *Archipetalia auriculata* and *Synthemiopsis gomphomac-romioides*, confined to certain Tasmanian waters, are likewise at risk.

Also at risk in Tasmania are various Trichoptera. *Taskiria mccubbini* and *Taskiropsyche lacustris* (Kokiriidae) and *Archeophylax vernalis* (Limnephilidae), all three of which occurred in Lake Pedder, may already be extinct; if not, they are certainly endangered species, for they have not so far been discovered in the new impoundment. *Westriplectes pedderensis*, another Pedder endemic, may possibly have survived the inundation, but in any event must also be regarded as an endangered species.

Finally among threatened insects, it is noted that the range of the Blephariceridae is diminishing because larvae are very intolerant to most forms of pollution and cannot survive in streams carrying sand or silt, or which have levels that alter as the result of upstream management of impoundment releases. Thus, *Edwardsina tasmaniensis* and *E. gigantea* appear to be seriously threatened. Indeed, *E. tasmaniensis* was thought to have become extinct until a recent rediscovery of it in southwest Tasmania.

All Tasmanian members of the Anaspidacea (syncarid crustaceans) are now officially recognized by the IUCN as vulnerable species, viz *Anaspides tasmaniae, A. spinulae, Paranaspides lacustris, Allanaspides helonomus* and *A. hickmani. Koonunga cursor*, found in Victoria, King Island and northern Tasmania, is under consideration for similar recognition. Probably all Australian anaspidaceans are vulnerable.

Astacopsis gouldi, the Tasmanian giant crayfish, is also officially recognized by the IUCN as a vulnerable species, but unlike the situation with Tasmanian syncarids, governmental conservation measures have been taken. To combat threats to this species by overexploitation as a food item, collections are now regulated and export is prohibited except by permit. To combat threats of habitat destruction, a small reserve, Caroline Creek, has been established. How effective these measures will prove remains to be seen. *Euastacus armatus*, the Murray crayfish, is another parastacid species that should be regarded as vulnerable. It is endemic to the River Murray system, where its numbers and distribution have declined markedly in the past century. Although still found in the upper Murray and its tributaries, very few have been sighted in South Australia during the past 25 years. Several other species of crayfish are probably threatened for many of the hundred or so endemic species of them, and some genera too, have small areas of distribution in regions increasingly affected by man. Genera with limited areas of distribution are *Engaewa* (extreme southwest of Western Australia), *Gramastacus* (Grampians, Victoria), and *Tenuibranchiurus* and *Euastacoides* (extreme southeast of Queensland).

Similar hazards face isolated populations of more widespread invertebrate species, and while the loss of these, if they show no differences from central populations, is not as important as the loss of isolated populations which do (as in the case of the stonefly species discussed above), it still represents a local decrease in species richness and a loss nonetheless. Some species inhabiting Tasmanian salt lakes provide examples. A recent survey of such lakes found the isopod *Haloniscus searlei* (Fig. 4.3(d)) in only two lakes, and the snail *Coxiella* in only one. The lake containing the largest population of *Haloniscus* (Township Lagoon, Tunbridge) is used partly as the local rubbish dump, and the lake with *Coxiella* (Folly's Lagoon, near Tunbridge) lies in an area liable to drainage for agricultural purposes (the adjoining lagoon, Clarke's Lagoon, has already been drained). Clearly these two species face extinction in Tasmania.

Knowledge concerning the conservation status of the Australian aquatic vertebrate fauna is on a *somewhat* firmer basis than it is for the invertebrate fauna — though that is not to say by any means that this knowledge is satisfactory.

Four fish species have been put forward as species whose existence is seriously threatened (Lake 1971): *Prototroctes maraena* (Australian grayling), *Macquaria australasica* (Macquarie perch), *Maccullochella macquariensis* (trout cod) and *Gadopsis marmoratus* (blackfish). In terms of the definitions adopted here, we may reasonably regard the trout cod as an endangered species since it is now known from only four localities. The other three are clearly vulnerable species — with the blackfish perhaps less vulnerable than the other two. *Prototroctes maraena* is further discussed in chapter 5. Six species have been put forward as fish whose distribution and/or abundance has been considerably reduced since the advent of European man to Australia: *Lovettia sealii* (Derwent whitebait), *Tandanus tandanus* (catfish), *Lates calcarifer* (silver barramundi), *Macquaria novemaculeata* (Australian bass), *Macquaria ambigua* (yellowbelly or callop), and *Maccullochella peeli* (Murray cod). Finally, there are numerous species (approximately 40) that are either not yet in a serious position from a conservation viewpoint but that are nevertheless threatened by contemporary changes to the nature of the aquatic environment, or that have restricted distributions; *Neoceratodus forsteri* (Australian lungfish) and most species of galaxiids fall into this category. Thus, although no freshwater fish species has apparently become extinct so far, there is no room for complacency.

Six species of frog are currently regarded as endangered: *Arenophryne rotunda, Litoria longirostris, Cophixalus saxatilis, C. concinnus, Rheobatrachus silus* and *Philoria frosti*. Their geographical distribution has already been indicated (Fig. 6.4). An important note here is that for the most part the endangered species inhabit montane areas, and, according to Tyler (1979), most lowland frogs are probably sufficiently widely distributed to withstand man's impact. Several states, nevertheless, have enacted legislation aimed at protecting frogs. The nature of threats to Australian frogs is discussed in chapter 6.

Four aquatic reptiles are of interest: two species of freshwater tortoise (*Pseudemydura umbrina, Carettochelys insculpta*) and the two crocodile species (*Crocodylus porosus, C. johnstoni*). The first of these, the western swamp tortoise, is threatened with extinction and is thus an endangered species. Its present natural population numbers less than 50 individuals, all restricted to two swamps near Perth. Both swamps are in nature reserves, but it seems likely that the species' days are numbered. The other, the pitted-shell turtle, is known in Australia from only a relatively small area of the Northern Territory, and on that account must be regarded as a rare species. It does, however, occur in Papua New Guinea. The estuarine and Johnston's crocodile are regarded as vulnerable species. Their current conservation status is discussed in a little more detail in chapter 6.

The conservation status of waterbird species is a matter of some debate. There are ornithologists whose opinion is that few species are seriously threatened at present. Others disagree. Whatever the case, over the years several species have been the subject of concern from conservationists. All species thought to be at risk are considered here. The species are *Cereopsis novaehollandiae* (Cape Barren goose), *Stictonetta naevosa* (freckled duck), *Tadorna radjah* (Burdekin duck), *Nettapus coromandelianus* (white pygmy goose), and *Malurus coronatus* (purple-crowned wren).

The Cape Barren goose occurs on offshore islands and coastal regions of southern Australia where it was regarded as an endangered species. Protection of breeding colonies — breeding birds were used as crayfish bait! — and other measures have undoubtedly improved the conservation status of this bird so that it no longer seems seriously threatened with extinction. The freckled duck is in a more parlous condition. The difficulty is that, despite total protection by game laws, many duck-shooters cannot or will not learn how to recognize the species; the result is that large numbers continue to be shot during each hunting season. Chapter 7 mentioned how in the 1980 season between 500 and 1000 specimens were shot at Bool Lagoon, South Australia. The species should certainly be accorded the status of vulnerable if not endangered. The same status should be given to the other two ducks, the Burdekin duck and the white pygmy goose. Both have probably been affected by reduction in available habitat. A similar process has given rise to the more hazardous position now occupied by the purple-crowned wren. For this species, however, it is not the diminution of suitable water-bodies, but more the destruction by cattle of the narrow and fragile strip of terrestrial vegetation (approximately 50 m wide) fringing the few (six) rivers where the purple-crowned wren still occurs. These rivers are all in northern Australia between latitudes 14 and 19°S: Kimberleys, Northern Territory, Queensland. The species has already been eliminated from the Ord and Leichhardt rivers. Despite total protection throughout its range, there seems little doubt that, unless suitable habitat is set aside, the conservation status of the species will move from endangered to extinct.

Finally it should be noted that all three native mammals closely associated with inland waters have been at one time or another the focus of conservation concern. Fortunately, none appears to be seriously at risk at present. The platypus is apparently secure, although it did approach extermination just 50 or so years after its discovery (chapter 8). Rigid protection undoubtedly saved it from annihilation by the fur trade. It continues to be fully protected. The fur trade also had a deleterious impact upon the water rat, together with other factors. This species, too, is now protected except in Tasmania where there is a short open season, and its conservation status can presently be regarded as secure; it is abundant in suitable habitats which are also abundant in all states except the Northern Territory, where, in any event, it is considered common. The false water rat, on the other hand, is a rare species; it is seldom collected and is known only from small areas within suitable habitats which are, however, common. It is a fully protected animal. More thorough collecting *may* reveal that it is more abundant than presently thought, as previously suggested in chapter 8. None of the other rodents characteristically associated with wetlands, though not strictly aquatic, seems to be threatened so far as current knowledge indicates.

The status of environments

Of course it makes no sense to consider conservation of the Australian aquatic biota without considering conservation of habitats; species and habitat conservation are mutually dependent. But the conservation of particular habitats has values over and above conservation for the biota within them. The conservation of habitats *and* biota with functional interdependence in a defined area, that is to say ecosystem conservation, has many values in addition to those directly attributable to plants and animals *per se*. These values are greatest, and most conspicuous, when the ecosystem in question is in a natural state or unmodified by man.

Throughout most of the world, the long-term preservation of natural ecosystems can only be ensured effectively by governments. Ecosystems must be set aside and protected by appropriate legislation. In Australia, they are protected within national parks, fauna and flora reserves, wildlife sanctuaries and in similarly named places. The total area of these in Australia, 4.1% of the total land area, is relatively small, despite the low human population density of the continent overall. Some states have very small percentages; Queensland has set aside only 1.7%.

Not all areas set aside as national parks and other protected areas were primarily designed to conserve natural ecosystems, or in fact do that satisfactorily, if at all. Additionally, whilst considerably more weight is now given to scientific and ecological criteria for establishing protected areas (and especially the need to preserve as much ecosystem diversity as possible), there are clearly many sorts of Australian aquatic ecosystems not yet adequately protected. This is exacerbated by the accelerating rate of alienation of environments throughout Australia in the name of 'progress'. There is, as it were, a set of endangered, vulnerable and rare aquatic ecosystems. No comprehensive inventory of these exists, though useful state-by-state surveys of wetlands have been made by the Commonwealth Scientific and Industrial Research Organization and various other bodies. The only conclusion to be drawn at present is that knowledge of the conservation status of natural aquatic ecosystems in Australia, on a comprehensive basis, is less than adequate, to put it mildly. This notwithstanding, it can be stated categorically that the future of several types of them is already in jeopardy. Endangered ecosystems include (the list is not complete) mound springs, rivers, marshes in temperate agricultural areas, swamps in the wet lowlands of Queensland, lagoons and other wet areas associated with northern rivers, River Murray billabongs, east coast dune lakes, and several aquatic sites listed internationally as waters proposed for conservation.

All of these ecosystems are significant elements of the Australian natural

Fig. 11.2 Some endangered aquatic ecosystems in Australia. (a) Coward Springs, a mound spring south of Lake Eyre; (b) marsh fed by spring near Chudleigh, northwestern Tasmania, a natural wetland in a temperate agricultural area; (c) lagoon near Mount Garnett, northeastern Queensland, a natural wetland in a semi-tropical agricultural area; (d) upper reaches of the Magela Creek, Arnhem Land, Northern Territory, a natural section of this northern river unaffected by buffalo activities.

landscape and have a unique combination of living and non-living matter. Our responsibility to prevent their extinction is as great as that to prevent the extinction of any Australian plant or animal species.

There are two main threats to mound springs (Fig. 11.2(a)): stock use and decreases in spring outflows. The effect of stock use is twofold. First, stock trample the littoral vegetation upon which much of the spring fauna depends (directly or indirectly), and secondly stock usually contaminate the water with excrement. The decrease in spring outflows is also attributable to stock, albeit indirectly. The problem is the recent fall in the water table resulting from excessive use of artesian and subartesian water for stock watering. Obvious protective measures are fencing off the springs and a more parsimonious utilization of underground waters. Further capping of bores would also help. Without such measures, the unique faunal assemblages present in Australian mound springs may soon be lost forever.

The conservation of rivers as ecosystems provides problems much less open to simple resolution. This is particularly because rivers represent an integration of many streams and their valleys; a river cannot be considered apart from its catchment basin. And, of course, many rivers themselves are dammed, support introduced species of plant or animal, or are otherwise polluted. Is it surprising, then, that only a small number of essentially natural rivers survive worldwide, and only a handful in Australia? Of those left here, most are faced by threats of one sort or another, or have already been subject to at least some modification (e.g. by the introduction of trout), despite location in sparsely populated regions (viz the Lower Gordon and Franklin Rivers in southwest Tasmania). The substantial modification of *all* natural rivers in temperate southern areas of Australia would be a particularly heavy loss in that they possess a large number of unique animals with southern hemisphere affinities and of great scientific interest.

A major contemporary threat to those remaining southern rivers that are still essentially natural is posed by damming, either to generate hydroelectricity, as in Tasmania, or to supply water, as in Victoria. With regard to damming, note that seemingly 'minor' hydrological changes can have profound repercussions upon the fauna. Even short-lived cessations of flow can extinguish whole groups; no blepharicerid larvae, for example, now occur in rivers below impoundments.

Intensive forestry practices (woodchipping) and land clearance for agriculture also pose threats to many southern rivers, as well as to many rivers elsewhere. Siltation caused by these practices drastically alters the fauna and is a long-term effect. Many engineers and others believe that if a little silt gets into a river it is flushed out in the next flood; but recent work has shown that this is not so and that silt penetrates deep into the interstices of the substratum and largely remains there after flushing. Many tropical northern rivers are threatened at present more by stock damage to banks and marginal vegetation than by damming or siltation. It is unfortunate that concern for the conservation of Australian rivers has lagged behind that for lakes, wetlands and other standing water-bodies.

The major threat to swamps and marshes (wetlands) in temperate agricultural areas (Fig. 11.2(b)) and the wet tropical lowlands of Queensland (Fig. 11.2(c)) is more obvious and total than any threat to aquatic ecosystems considered thus far. It is drainage to provide agricultural land and mitigate flooding. In this respect, most landowners are quite unaware of (or choose to ignore) any responsibility for the conservation of Australian wetlands, and act not as exploiters *and* custodians of the land, but exploiters alone. As a result, only a relatively small number of natural swamps and marshes remain in many agricultural areas.

Wetlands in the north of Australia face threats of a different sort. Many in the northern swamp lands of the Northern Territory have been severely damaged by the introduced water buffalo. As chapter 8 indicates, these often overgraze, wallow excessively, and generally modify the natural character of lagoons, swamps and rivers of the region in an unfavourable way (Fig. 8.5). Over a wider area of northern Australia, cattle cause damage by trampling and otherwise affecting the fragile but important strip of terrestrial vegetation fringing the banks of rivers and streams (Fig. 11.2(d)). The catastrophic effect of this upon populations of the purple-crowned wren has been noted.

On a global scale, the decreasing area of wetlands is of international concern, and several programmes to aid wetland conservation are in progress under the aegis of the United Nations and other international bodies (e.g. IUCN). Waterfowl represent an element of swamp and marsh ecosystems most noticeably affected by the loss of wetlands, but wetlands have many values in addition to that as a waterfowl habitat: they act as natural hydrological sponges, as sinks and filters for catchment material, and as nutrient sources downstream.

Many threats face billabongs associated with the River Murray (Fig. 11.3(a)). Impoundment and flow regulation of the river itself are important ones, but others include drainage, and use as sewage lagoons, garbage dumps, and water storages. Few if any of the billabongs of the upper Murray (near the Hume impoundment) are in a pristine condition, yet the region used to be characterized by a multitude of billabongs and swamps associated with the river. For most of its length, the Murray flows through semi-arid regions and the significance of its floodplain waters to a wide variety of both terrestrial and aquatic plants and animals is profound. Yet national parks bordering its lowland reaches are small, few, and far between. South Australia, whose economic existence depends upon the river, has none.

Fortunately, some of the coastal dune lakes of eastern Australia (Fig. 11.3(b)) are within national parks. However, they are particularly susceptible

Fig. 11.3 More endangered aquatic ecosystems in Australia. (a) Billabong associated with River Murray, near Albury; (b) coastal dune lake near Cape Bedford, northeastern Queensland.

to damage (they drain if the lake bottom is breached), not all are identical, and many have already disappeared. As fresh but coastal bodies of water with some unique species, it is not surprising that they have attracted considerable attention from freshwater biologists. Sand mining is a major threat, and no amount of repair and restoration can regenerate them once destroyed.

Table 11.2 Present conservation status and main threats to Australian sites listed in *Project Aqua* as waters proposed for conservation. Note that no rivers were proposed for conservation by *Project Aqua*!

Water-body	State	Type	Conservation status	Main threats
1 Lake Tarli Karng	Vic.	Highland lake	Secure	None
2 Anabranch lakes of Hattah Lake System	Vic.	Anabranch lakes of R. Murray	Vulnerable	Forestry, flow regulation
3 Lake Gnotuk*	Vic.	Deep saline lake	Vulnerable	Catchment alteration
4 Lake Eyre	SA	Ephemeral salt lake	Secure	None
5 Salt lakes between Robe and Beachport	SA	Ephemeral/semi-permanent salt lakes	Vulnerable	Drainage
6 Lake Corangamite	Vic.	Permanent salt lake	Vulnerable	Flood control
7 Red Rock Complex of Lakes*	Vic.	Fresh to salt lake complex	Endangered	Catchment alteration
8 Little Llangothlin Lagoon	NSW	Wetland	Endangered	Drainage
9 Lakes Hiawatha and Minnie Water	NSW	Coastal dune lakes	Secure	None
10 The Gap Lagoons	NSW	Coastal dune lakes	Vulnerable	Mining, catchment alteration
11 Lakes Eacham, Barrine Euramoo* and The Crater	Qld	Deep volcanic, fresh lakes	Secure	None
12 Lake Blue, Cootapatamba, Albina, Hedley Tarn, Club	NSW	Glacial lakes	Secure	None
13 Lake Seal	Tas.	Glacial lake	Secure	None
14 Lake St Clair*	Tas.	Glacial lake	Vulnerable	Water level change
15 Lakes Sorell and Crescent	Tas.	Freshwater lakes	Vulnerable	Use as reservoir
16 Lagoon of Islands	Tas.	Wetland	(Flooded)	—
17 Lake Pedder	Tas.	Freshwater lake	(Flooded)	—
18 Lake Judd	Tas.	Glacial lake	Secure	None
19 Lake Fergus or Lake Ada	Tas.	Freshwater lake	Vulnerable	Hydro-Electric Commission
20 Macquarie Island Lakes	Tas.	Freshwater lakes	Secure	None

* See Fig. 11.4.

Some of the sorts of threatened aquatic ecosystem mentioned in preceding paragraphs were listed as named Australian sites in 1971 in *Project Aqua*, an attempt by the International Biological Programme to inventory the world's most significant inland waters proposed for conservation. The list of Australian sites is reproduced in Table 11.2, which also indicates their present conservation status. A few are illustrated in Fig. 11.4(a)–(d). Not all sites mentioned by *Project Aqua* fall within the endangered category, but certainly several are threatened and two important ones, Lake Pedder and the Lagoon of Islands (both in Tasmania), have been lost since 1971. Lake Pedder (Fig. 11.1) was the only lake of its type in Australia and had several species considered endemic to it at the time of their discovery. The lake was flooded by the Serpentine-Huon impoundment created by the Hydro-Electric Commission of Tasmania — despite being in a national park! And then, as if to seal its fate, the Inland Fisheries Commission introduced several thousand trout fingerlings into the new impoundment. The Lagoon of Islands, also the only one of its type in Australia, was characterized by floating islands of vegetation formed by a unique process of ecological succession. It was destroyed as a natural ecosystem by the Hydro-Electric and Inland Fisheries Commissions of Tasmania. Considered as a lake of outstanding scientific interest, its destruction was brought about by raising its level.

Fig. 11.4 Some Australian lakes listed by *Project Aqua*. (a) Red Rock Complex of lakes, Victoria; (b) Lake Gnotuk, Victoria; (c) Lake Euramoo, Queensland; (d) Lake St Clair, Tasmania.

Of endangered Australian ecosystems amongst those listed by *Project Aqua* (see Table 11.2), special mention should be accorded the Red Rock Complex of lakes in western Victoria (Fig 11.4(a)). This set of eight small lakes displays salinities from fresh to salt and thus provides a unique opportunity for limnological studies. However, their entire catchment is grazed, subject to erosion and part ploughed, and a causeway has been constructed to the only island present. Recommendations to the Victorian Government that they be given a high level of protection in view of their importance have recently been accepted. All that remains is for government policy to be implemented! At present, it is not.

Finally, with respect to Australian lakes listed by *Project Aqua,* it should be stressed that the omission of many of the aquatic ecosystems considered in this chapter as worthy of conservation does *not* reflect any lowered priority or lack of merit in those not listed. It merely reflects the extent to which *Project Aqua* needs considerable revision so far as Australia is concerned.

Overall, this consideration of the conservation status of Australian aquatic ecosystems is a distressing indictment of our lack of a national water resource policy (and the necessary legislation to implement it) designed to protect aquatic ecosystems as an integral part of our scant but vital water resources.

12   The impact of man

The previous chapter noted that man's impact upon aquatic ecosystems has been profound. This has certainly been the case globally, but the statement applies with equal force to Australia: despite our low ratio of people to land (2 persons/km²), the relatively recent origin of significant impacts (Caucasian not Aboriginal man is to blame), and our essentially urban demography, few Australian lakes, rivers and other inland waters remain in natural condition. For well-watered regions, the only major exceptions occur in southwestern Tasmania and parts of northern Australia (e.g. the Kimberleys). Remote and arid areas also still have some natural waters, though, of course, these are few, generally ephemeral, and scattered.

The present chapter examines the nature of man's impacts and their broad effects.

INTRODUCED BIOTA

Several species of introduced animals and plants are known to have had significant effects, and others are assumed to have had an impact, though there is little firm evidence of its precise nature. Three general impacts seem likely from introduced animals: habitat alteration, competition with native animals, and predation upon them. Animals known to be important in this respect are the fox, water buffalo, cane toad and trout. Probably important are the European carp and mosquito fish.

The major effect of foxes is bankside predation upon vertebrates. On the River Murray in South Australia, for example, foxes destroy over 90% of freshwater turtle eggs. The effect upon waterbird eggs is likely to be just as profound. The water buffalo, as indicated previously, causes severe damage to water-bodies in the far north wherever large populations build up. The effect of the cane toad is less obvious, but when large populations of this species develop, fetid conditions are often created. The species has probably also had a negative effect on native amphibians, and in any event is a nuisance and pest throughout most of its range. There is good evidence that trout (*Salmo trutta*) competes with galaxiid fish (see chapter 5), and circumstantial evidence that it does so with others and is a significant predator of certain invertebrates (e.g. *Anaspides*). The impact of European carp has not yet been unequivocally determined, but in the opinion of many biologists it destroys aquatic plants by uprooting them (Fig. 12.1) and makes waters turbid by disturbing bottom muds. The impact of the mosquito fish, likewise, has not yet been unequivocally determined, but it seems likely that it is an important predator of many invertebrates.

Two further and more specific impacts resulting from animal introductions merit brief note, although the actual importance is not yet clear. First, the native black duck (*Anas superciliosa*) is thought to be in danger through interbreeding with introduced and feral mallards (*Anas platyrhynchos*). Second, the importation of large numbers of aquarium fish (see chapter 5) is thought likely to have introduced some exotic fish diseases and parasites.

Of introduced plants, those with most environmental impact are the water hyacinth (*Eichhornia crassipes*), salvinia (*Salvinia molesta*), alligator weed (*Alternanthera philoxeroides*) and Canadian pond weed (*Elodea canadensis*). The major effect of these is to clog water-bodies with excessive growth. Chapter 10 discusses this in more detail.

Fig. 12.1 Uprooted *Typha*, Lake Burley Griffin, Canberra. Ecologists studying this lake claim such damage is caused by the activity of carp.

POLLUTION

Many define water pollution as *any* change in the natural character of a water resulting from man's activities. The presence of introduced animals would be regarded as a form of pollution in this definition. For present purposes, however, a narrower definition is adopted, namely, that *water pollution is a significant and deleterious change in the natural character of a water resulting from the addition of material or heat by man.* Particular changes include the creation of health hazards, unsightly appearance, unpleasant smells, murkiness, excessive plant growth, undesirable physical conditions, and many other features. In short, then, pollution involves a degradation in the value of water as a resource.

Many sorts of material are discharged into inland waters by contemporary man and cause pollution. Their nature and effects, however, fall into relatively few categories.

Non-poisonous organic pollutants

Organic pollutants that are not directly toxic derive from two main (but not the only) sources: domestic households and certain industries. Sewage and detergents are the two principal types of domestic waste involved. Most of the former comprises faecal material and urine and is usually treated by municipal or other authorities before discharge as sewage effluent. Although effluent disposal to land and the sea is important, much sewage effluent is also discharged into inland waters. The aim of treatment is to degrade the complex organic materials contained in the waste to simpler organic and inorganic materials, and render harmless any pathogenic bacteria, parasite eggs or other disease-producing materials that may have been present in the 'raw' sewage. Detergents contain both organic material and plant nutrients (phosphorus) and, unlike sewage, are sometimes not easily biodegraded. That, at least, is so with 'hard' detergents, those with a branched carbon chain molecular structure. 'Soft' detergents, those with linear carbon chain molecules, are biodegradable, and for that reason are replacing 'hard' detergents.

Non-toxic organic trade wastes emanate particularly from food industries: breweries, dairies, abattoirs and canning factories are all important. Other contributors include papermills, tanneries and petrochemical factories. Effluents from places where intensive animal husbandry occurs, e.g. piggeries, may also be mentioned here. The nature of wastes varies considerably, as does its stability in the environment: as a rule, animal wastes or wastes from food factories are quickly degraded, but paper mill wastes may be refractory or at least fairly resistant to breakdown. Carbohydrates, proteins, fats, amino acids, cellulose and wood sugars are the main types of organic compound involved, but petrochemical wastes, oil, and other types of organic waste may be locally important.

The primary and most significant effect of non-toxic organic wastes on inland waters is deoxygenation. This is mainly the result of bacterial breakdown of the waste. The intensity of deoxygenation relates to many factors, but temperature, the nature and quantity of the organic waste, and the occurrence of suitable bacteria are particularly important. In simple terms, the bacteria degrade the complex organic molecules of the waste first to less complex molecules, and ultimately to inorganic materials. The process is aerobic, thereby removing oxygen from the water. If the pollutant load is heavy, all oxygen disappears and anaerobic conditions follow (oxygen completely absent).

In rivers, the major type of inland water affected, reoxygenation occurs downstream as a result of photosynthesis and atmospheric diffusion, the rate of the latter being governed by turbulence, temperature, the presence of oil films and detergents, and the extent to which the river is deoxygenated. The characteristic pattern of oxygen concentration downstream from a single source of pollution is referred to as the 'oxygen sag curve' when plotted graphically (Fig. 12.2).

Many important biological events follow the lowering of oxygen levels. An obvious one is the disappearance of the normal fauna and flora, since most aquatic life cannot stand conditions of lowered oxygen concentrations for any length of time. They are replaced by a less diverse biota more tolerant to low oxygen conditions and the composition of which is determined by the severity

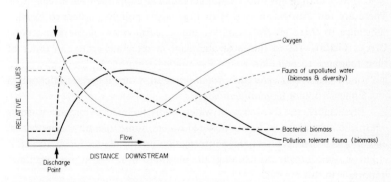

Fig. 12.2 Biological relationships and oxygen gradients in a hypothetical stream receiving organic waste at a single location.

of pollution, the nature of the environment and geographical factors. In simple situations in rivers (point source pollution), the major elements of this
replacement biota exhibit a longitudinal downstream sequence or pattern correlated with the form of the oxygen sag curve (Fig. 12.2). Such simple situations, however, are increasingly rare, and the more normal one is where pollution increases in intensity downstream, with no recovery possible.

With regard to the 'flora' of organically polluted rivers, note that bacterial
numbers are highest immediately after effluent discharge and decrease
downstream as the waste is mineralized. Except in the most grossly polluted
parts of a river where there is no oxygen present at all, a more obvious element
of the 'flora' of badly polluted reaches is 'sewage fungus', that is, a characteristic
assemblage of bacteria, fungi and protozoans. Frequently, sewage fungus
forms large unsightly ragged masses from white to brown in colour, and cover-
ing large areas of the bottom. The principal organism involved is *Sphaerotilus
natans*, a sheathed bacterium, but other bacteria, many true fungi, and
various ciliate protozoans are important as well. With oxygenation
downstream, various algae become the dominant plants, often exhibiting a
more or less predictable successional pattern.

Few large invertebrates occur in conditions of severe deoxygenation, and
those that do obtain oxygen directly from the air by special means such as
breathing tubes. Larvae of the mothfly, rat-tailed maggot and certain mosquito species are characteristic here. When at least some oxygen is present, the
most characteristic animals are small oligochaete worms of the family
Tubificidae. Enormous populations of these may build up. With further increase in oxygen concentrations, they are usually replaced by red chironomid
insect larvae, so-called 'blood worms'. These, too, may build up large populations. Further improvement in oxygen conditions leads to an increased diversity of species until finally the invertebrate fauna of unpolluted water reappears.
There are few Australian studies of organically polluted waters to provide
substance to these general remarks, but a recent one—of the Dandenong
Creek, Victoria—listed the most abundant invertebrate groups in order of
their tolerance to pollution as the Psychodidae (mothflies), tubificid worms,
gastropod snails (*Isidorella* then *Potamopyrgus*), Chironomidae, Trichoptera
and Ephemeroptera (*Atalophlebioides*).

Predictably, the fish fauna of waters badly polluted by organic waste is
conspicuous by absence. In less polluted waters, some tolerant species occur,
and although there is little precise information on this matter for Australia, it
is known that introduced fish (e.g. the mosquito fish, roach) are sometimes
important in this respect.

In summary, non-toxic organic waste (a) decreases the number of species
below naturally occurring numbers, (b) leads to the dominance of the com-

munity by just a few species, and (c) changes the aquatic biomass present (the amount of living material) so that in severely polluted situations it is low, but in mildly polluted situations is very high. Additionally, two features characterize rivers: first, they *can* recover from the effects of waste (given the opportunity), and, second, in simple situations their biota and some physicochemical conditions exhibit a longitudinal pattern.

Finally, with regard to the effects of non-toxic organic wastes, it should be added that effects are not restricted to an impact upon oxygen concentrations and the consequences of that. Almost always associated are effects resulting from the addition of particulate matter and detergents over and above contributions to the organic loading. Thus even mild turbulence may cause foaming if detergents are present.

Plant nutrients

Sewage effluents and detergents, in addition to their organic content, contain many plant nutrients, particularly phosphorus and nitrogen. Urban run-off, garbage tips, certain industries, soil erosion, motor vehicles, fertilized agricultural areas and even atmospheric fallout are also contributors of plant nutrients to inland waters. The end result, often, is that unnaturally large quantities enter many inland water-bodies. The corollary is predictable. Given favourable light and temperature conditions, plant growth is stimulated, sometimes excessively so. In broad terms this phenomenon is referred to as *eutrophication*, or more specifically, *cultural eutrophication*.

Both large plants (macrophytes) and microscopic free-floating or attached algae (phytoplankton or phytobenthos) are involved. In either case, excessive growth causes problems. Too many macrophytes clog up waterways, decrease recreational usage, and may lead to increased numbers of aquatic insect pests. Too many algae may discolor the water, be toxic, smell, clog filters if the water is used as a source of domestic supplies, form floating putrescible masses or slimy coatings to rocks, or be otherwise displeasing aesthetically. Associated with these changes in plant biomass may be others. There may be a change in the nature of the fish fauna, for example, or deoxygenation may result from the decay of large amounts of plant material.

All in all, eutrophication is a process which tends to degrade the natural character of lakes (as well as giving problems in impoundments), and its causes, excess plant nutrients, in this sense constitute pollutants. Strategies to prevent eutrophication are the restriction of nutrient input to water-bodies, and nutrient removal from them. The former strategy includes the diversion of sewage effluents, and the removal of nutrients from sewage and detergents by treatment. Harvesting, flushing, inactivation and the treatment of sediments are some techniques for removing nutrients from water-bodies. The symptoms of eutrophication are also amenable to control; macrophytes, for example, can be kept from becoming a nuisance by periodic cutting or cropping. With suitable management, then, the eutrophication of standing waters is not inevitable and the process can be controlled. Again, there are strategies that may be able to restore pre-eutrophic conditions to lakes that have been allowed to become eutrophic, though they are not always easily applied.

Eutrophication threatens several Australian lakes and impoundments of which Lake Burley Griffin, an artificial lake in the nation's capital, is perhaps the most notable.

Inert material

More or less chemically inert waste constitutes another class of pollutant. It usually takes the form of finely divided particulate matter or insoluble liquids, but large refuse items such as worn tyres, wrecked cars, plastic containers and the like are a not insignificant form. Sources of fine material are paper mills producing wood pulp and fibrous waste and mines producing tailings. Whatever the source of fine material, the effect is profound; lake and river bottoms are blanketed, light penetration is decreased, and animal and plant biomass falls markedly. Large refuse items at the very least are eyesores (Fig. 12.3).

Poisons

Many products of man's industrial, agricultural, urban or other activities enter inland waters and behave as poisons. Some do this because they are manufactured poisons used initially against non-aquatic forms of life; others are toxic, in a sense, incidentally. But, toxic by design or accident, the general effect is the same. The overall ecological effect is to decrease both the diversity and abundance of aquatic life. And if allowed to contaminate waters used for drinking purposes, the effect on man may be acute or chronic poisoning, the development of cancers, or an increase in the number of fetal malformations.

The extent of ecological impact is determined by many factors. One, of course, is the chemical structure of the poison in question. Others are concentration and dosage, exposure time, the presence of other poisons and solutes, water temperature and salinity, and oxygen concentration. Toxicity testing, that is, the assessment of the degree to which a given compound is poisonous

Fig. 12.3 Public abuse of an urban creek, Melbourne.

to aquatic life, has developed into an important branch of freshwater biology. However, results gained elsewhere must be applied with caution to Australian species.

Both inorganic and organic compounds are implicated as aquatic poisons. Toxic inorganic compounds include acids, alkalis, sulphides, cyanides and salts of arsenic, lead, chromium, zinc, cadmium and other heavy metals. Special mention may be accorded cadmium for it now appears that it is one of the more toxic heavy metals yet has been overlooked in importance until recently. Chemical factories, certain engineering plants such as electroplating works, and mines are frequent sources of these wastes. In the case of mining, wastes often continue to be added to the natural environment long after mining operations have ceased. Thus, zinc and other heavy metals were added to the River Molongo in the Australian Capital Territory for many years after the mine at Captain's Flat upstream had closed. Likewise, mercury persists in the sediments of the Lerderberg River, Victoria, although gold mining, which used mercury in the extraction process, ceased some 50 years ago. Remedial measures against mining wastes are almost always costly and often not very effective.

Organic poisons are many and varied. Important ones are pesticides, herbicides, polychlorinated biphenyls (PCBs), phthalate esters, phenolic compounds, halomethanes, and various heavy metals bound to organic molecules (such as methyl mercury). Only pesticides and herbicides are used directly by man as poisons, and then usually on land. Spray contamination, run-off and atmospheric fallout soon spread them to the aquatic environment, however. The major sorts of pesticides are chlorinated hydrocarbons (e.g. DDT), organo-phosphorus compounds (e.g. Malathion, Parathion), and carbamates (e.g. Carbaryl).

Ecologically, three general features of herbicides and pesticides in the aquatic environments are noteworthy. First, they may be very toxic to aquatic life, with only minute concentrations needed to poison living organisms. Second, some may persist in the environment for long periods. And third, they may accumulate and concentrate in aquatic food chains. Less is known of the ecological effects of other organic poisons, though enough is known about their effects on man to give rise to considerable concern when they occur in drinking water supplies.

Heat

Thermal pollution, the elevation of natural water temperatures by the addition of heated effluents, is most frequently caused by the use of river water to cool electric power generating plants. In Victoria, such pollution occurs in the Latrobe River which is used by the State Electricity Commission to provide coolant (Fig. 12.4). The extent to which the temperature of this river may be raised is subject to strict control. High water temperatures may kill the biota in whole or part, lower oxygen concentrations yet increase respiration rates, and have many indirect effects.

Radioactive wastes

Low energy radioactive wastes may be discharged to rivers or temporarily stored in impoundments by mining and certain industries, atomic energy research establishments, hospitals and universities. The amounts are much more strictly controlled in Australia than is the case with other potentially dangerous pollutants. Two features of the ecological impact of the waste are important: the radioactivity may be concentrated by the aquatic biota in the

Fig. 12.4   Discharge into the Latrobe River of heated water from a thermal electric generating plant.

food chain in the same way as some pesticides are, and there may be a differential accumulation of radionuclides by uptake organisms.

CATCHMENT CHANGES

Almost all inland waters reflect events that occur on catchments. The only major exceptions are springs. As a result, many of man's activities, though physically removed from water, ultimately affect inland waters if in some way they alter the nature or pattern of catchment run-off.

Deforestation, overgrazing and afforestation are three activities with important catchment repercussions. With deforestation, run-off may be greater in quantity, have a significantly altered seasonal pattern, and contain more salts, plant nutrients and silt. Similar results obtain when catchments are overgrazed and thus eroded. In the upper reaches of streams and rivers, the end result may be that clear permanent waterways with low concentrations of salts and plant nutrients become turbid, silted-up and ephemeral, and have higher salinities and concentrations of plant nutrients. Afforestation does not have converse results, however, since for the most part in Australia afforestation involves the planting of conifers on land previously cleared of native trees. Indeed, afforestation of Australian catchments with conifers has its own deleterious effects, though these have yet to be fully documented. The loss of direct inputs of natural plant material to small streams is certainly an important one.

Whether planted or natural, the burning of forests has an important effect on catchment waters. After a fire, a major result is that streams and rivers in the catchment inevitably carry higher sediment loads; this also happens on a more local basis when logging, roadmaking and other activities occur that disturb the vegetation cover and soil of forests. Of course, forest fires occurred before man came to Australia, but at least in temperate forests probably had a different pattern; they were, perhaps, more frequent but less severe.

Irrigation is another catchment activity of interest here. One of its important effects in parts of southeastern Australia is to elevate significantly river and stream salinities, effects of considerable concern with particular regard to the River Murray. In irrigated parts of the Murray valley, the problem mainly results from over-irrigation which causes the underlying saline groundwater to rise. Capillary action and evaporation then lead to soil salinization, and eventually to the discharge of saline effluents to the river. The cultivation of non-irrigated crops (dryland farming), too, usually has an impact — an impact analogous to, if not as pronounced as, that of overgrazing and deforestation.

Finally, it should not be forgotten that man's houses and factories, in addition to being direct sources of pollutants, further affect inland waters as part of urban catchments. Run-off from such catchments is greater in total quantity than natural catchments, direct flows are large with high rates, and suspended sediment and plant nutrient loads are frequently significant.

EXPLOITATION OF AQUATIC ANIMALS

Several animals living in Australian inland waters are of use to man in so far as they can provide food, fur or leather, or because their pursuit provides sport. Exploitation of some of them has undoubtedly affected the size of populations, and, indeed, as the last chapter indicated, has placed some of the species involved in a threatened position from the viewpoint of conservation. That was certainly so, for example, with the platypus which was exploited for its fur, and freshwater crocodiles exploited for their hides.

At present the most significant impacts that exploitation of this sort seem most likely to have concern some fish and duck populations. Thus, the present small populations of the Tasmanian whitebait (*Lovettia seali*) reflect recent commercial overfishing (now stopped). Overfishing has also been put forward to explain the recent decline of some populations of the Macquarie perch (*Macquaria australasica*). The impact of duck-shooting, apart from that on a few threatened species, is difficult to assess. However, it can hardly be negligible given the popularity of the sport in Australia, and the size of at least some hunters' bags.

DIRECT HYDROLOGICAL IMPACTS

Impacts discussed so far largely concern either the animals or plants of inland waters or the physicochemical nature of their environment. But several of man's impacts go much further. He has created many new water-bodies, and drained or fundamentally altered others.

Three main types of artificial water-body can be distinguished: small farm dams (or ponds), large reservoirs (or impoundments), and water supply canals. So many new water-bodies have been created that a basic alteration has been effected in the nature and distribution of Australian inland waters (see also chapter 2). In particular, permanent and often deep bodies of fresh water are now common throughout most of the temperate parts of Australia, and small and usually permanent freshwater-bodies occur in all but the most arid regions (Fig. 12.5). Likewise, in some semi-arid regions, numerous artificial canals carry water long distances for purposes of irrigation and for stock and domestic supplies. In the Snowy Mountains a complex system of reservoirs, canals and tunnels has even reversed the direction of natural flow of part of the region in order to augment inputs to the River Murray (Fig. 12.6).

Although there are several notable examples of natural lakes that have been flooded or had their levels raised by damming (Lake Pedder, Lagoon of Islands, The Great Lake, Lake St Clair), the major direct impact of most Australian reservoirs is upon river systems. The effect is always profound, though its full extent is only just beginning to be appreciated. Physical barriers to animal migrations — particularly of fish — are created. Temperature and oxygen regimes are altered. Seasonal patterns of flow are often reversed. And fundamental changes to the nature and abundance of riverine animals and plants result.

Many examples can be provided. The distribution of the barramundi, *Lates calcarifer*, is being restricted by impoundments in northern Australia (see chapter 5). In the River Murray downstream of Hume Dam, the natural amplitude of water temperature is decreased and its seasonal pattern altered, and oxygen concentrations immediately below the dam may become unnaturally low. The general effect of the Hume impoundment on River Murray flows is to reduce seasonal values and maintain unnaturally high ones over the summer. Considering the river as a whole, the decline of at least some native fish can be attributed to the effects of impoundments. Further, the

Fig. 12.5 Farm dam in the Gawler Ranges, a semi-arid area of South Australia. The birds are hoary-headed grebes.

Fig. 12.6 Lake Eucumbene, NSW. The impoundment is the main storage reservoir of the Snowy Mountains Scheme. Completed in 1958, it holds up to 4.8 km^3 of water, and its dam wall is 116 m high and 0.8 km wide at its base. Water from this impoundment passes in tunnels to regulate storages in other impoundments of the scheme. Australian Information Service photograph.

zooplankton is now that characteristic of standing, not running, waters (by contrast, the zooplankton of the Darling River, a largely unimpounded river, is still typically riverine in essential features). Again, *Alathyria jacksoni*, a riverine species of freshwater mussel in the River Murray, is gradually being replaced by *Velesunio ambiguus*, a species more typical of billabongs, wetlands and other standing water-bodies.

The reverse of the coin, so to speak, is that man has drained many natural lakes and wetlands. The extent of wetland drainage has already been referred to with alarm in Chapter 11.

Finally, for rivers, note that many changes to their physical nature have been made under the name of 'river improvement schemes'. Channels are straightened, banks cleared of natural vegetation, bottoms desnagged, and in general attempts are made to make the river behave more as conduit than as an integral part of the natural environment.

13 Records from the past

Lakes respond sensitively and promptly to climatic, geological and biological events that operate directly on them and their catchments. Lake sediments integrate these responses, with the deepest sediments recording the earliest events and the most shallow, the most recent ones. Thus, lake sediments represent an historical record of past climates and environments.

The study of past records in lake sediments is known as palaeolimnology. Palaeolimnology, then, is that branch of limnology concerned with the origin and geomorphological history of lakes, and of their responses to climate and the changing inputs of waters and of dissolved and suspended materials. Many North American and European lakes have been studied, but it is only recently that rigorous palaeolimnological studies have commenced in Australia. Already, however, they have yielded exciting and important information, and promise much more. At the same time, those who study past and modern Australian lakes seem to be less aware of each other's work than is the case in North America and Europe. This chapter is a partial response to that phenomenon. It represents an effort to draw the attention of the wider readership of a book like this to the importance of palaeolimnological investigations within the total framework of studies of inland bodies of water. As such, no attempt is made to review Australian palaeolimnological studies (such presumption would serve only to alienate Australian palaeolimnologists!). Rather, the chapter aims to summarize the broad nature of the records contained in lake sediments, and to illustrate this by Australian examples.

LAKE AGES

Most lakes are geologically young, and palaeolimnology is therefore usually concerned with a time-scale of less than 25 000 years. Many Australian lakes, in fact, are less than 10 000 years old. Nevertheless, one of the most exciting findings of recent palaeolimnological studies is the discovery that many Australian lakes are of great antiquity. Some have histories that extend back *millions* of years. The saline sediments of Lake Tyrrell, Victoria, for example, are some 400–500 000 years old and postdate an older lake (Lake Bungunnia) which extended back 2.5 million years. Lake George, near Canberra, was in existence as an active lake in late Miocene times (8–10 million years ago). Lakes Eyre and Frome are the remnants of Miocene lakes in central Australia dating from at least 15–20 million years ago. And elongated playa lakes in Western Australia occupy the sites of old drainage lines of Tertiary age (between 2 and 70 million years ago).

With these ages, some Australian lakes are even older than the well-known ancient lakes of East Africa or Lake Baikal. However, it is important to note that, unlike these permanent freshwater lakes, ancient Australian lakes were temporary-water bodies; their records indicate considerable ephemerality, and often the presence of saline water.

Records from other less ancient lake sediments in Australia are also of considerable interest for many of these, too, are older than 25 000 years. Records from the site of Lynch's crater (now a drained swamp) in northeastern Queensland extend over the past 150 000 years. Mowbray Swamp in Tasmania has sediments spanning the past 110 000 years, and the nearby Pulbeena Swamp, the past 80 000 years. Lake Leake in South Australia appears to be about 50 000 years old.

The age of lakes, and more important the age of deposition of a given stratum of sediment, is determined by a variety of dating techniques. Radiometric techniques, those involving the determination of the concentration of certain radioactive isotopes, are especially important.

The most useful technique involves a carbon isotope, C_{14}, and can be used to date sediments between 150 and about 70 000 years of age. Its methodology involves the determination of the ratio of a stable to an unstable carbon isotope, viz C_{12} to C_{14}. Unfortunately, the technique is not useful when sediments are contaminated by allochthonous carbon, as when charcoal derived from human activity or sediments of very recent origin are mixed with older material. In the latter case, the worldwide combustion of fossil fuels of the last two centuries has altered the previously constant ratio of carbon isotopes in the atmosphere. With the establishment of carbon dating laboratory facilities in Australia, almost all Australian palaeolimnological studies now involve the age determination of reference points in sediment cores using the C_{14} method.

To date sediments too young for the application of the C_{14} method, two other radiometric techniques have been used. One involves a radioisotope of caesium, Cs_{137}. This is released into the atmosphere by above-ground nuclear explosions, and thus appeared first in sediments (the isotope is washed from the atmosphere by rain) about 1954, with concentrations peaking between 1959 and 1963. It is relatively immobile in sediments and hence provides a unique datum for very recent strata. The other method to date recent sediments involves a lead isotope, Pb_{210}. The method is a relatively new one and at present its results are somewhat controversial. Contemporary radiocarbon levels in swamp sediments, since the atom bomb tests began in 1950, have also been used to date sediments over the last 30 years.

Other radiometric techniques are available for dating sediments too old for the application of the C_{14} method. They involve the determination of the ratios of isotopes of uranium and lead ($U_{238}:Pb_{206}$, $U_{235}:Pb_{207}$), rubidium and strontium ($Rb_{87}:Sr_{87}$), and potassium and argon ($K_{40}:A_{40}$). The potassium/argon method is particularly useful. Isotopic geochemistry also has a role to play in palaeolimnology other than as a dating tool, for isotopic fractionation processes may provide information on palaeoenvironments. The proportions of O_{16} and O_{18} in calcareous fossils provide a means of estimating temperature at the time the animal lived. And the proportions of C_{14} to C_{12}, and of S_{34} to S_{32} may give clues on lake metabolism and microbial activity.

Two non-radiometric dating techniques should also be briefly mentioned. One involves the determination of the compass direction of iron particles in sediments (remanent magnetism); the palaeomagnetic direction reflects the polarity of the particles when deposited, and this can be equated fairly precisely with former magnetic deflections. The method has been applied recently in a number of lakes in southern Australia. In a few cases, seasonally different layers in sediments can also be used as a dating technique.

SEDIMENTARY RECORDS

Lake sediments have numerous sources, but are greatly influenced by the geology and topography of the lake and its catchment. The basic material essentially results from the interaction of local climate and geology, but of course is frequently much modified by conditions in the lake. Thus, wave action and currents may rearrange material into coarser or finer particles, and resuspension and redeposition of material is an ever present difficulty for all palaeolimnologists. A variety of materials, both inorganic and organic, are added to this basic material and reflect conditions in the lake at the time of their addition. All of these sedimented materials provide many records that document the history of the lake.

Though considerable care is needed in interpretation, much can be learnt of past conditions merely from examining the physical structure of sediments. Thus, it is sometimes possible to make direct estimates not only of age but also sedimentation rates from examination of sediment layers. Sedimentation rates closely reflect catchment erosion rates, a phenomenon which itself may be subject to further explanation. Unfortunately, relatively few lakes show clearly layered sediments, but gross sediment composition can nevertheless yield useful information. Volcanic ash, for example, in the sediments of some western Victorian salt lakes clearly documents the former occurrence of at least nearby vulcanicity, and sedimentological analyses alone enabled the construction of a chronology of major limnological events in Lake Keilambete, Victoria. Saline and freshwater phases were clearly indicated, as well as low and high water levels. Details of the structure of a single core of the sediment from this lake are indicated in Fig. 13.1 to illustrate the sort of variation that occurs according to depth (and hence time).

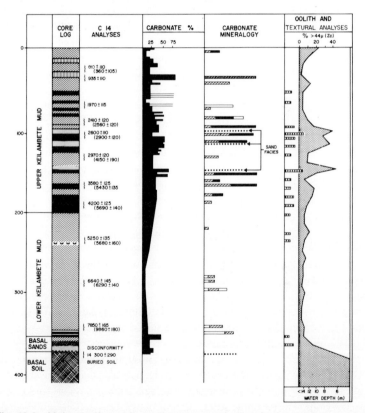

Fig. 13.1 Main stratigraphic, sedimentary and other features of a core from Lake Keilambete, Victoria. From Bowler (1981). Core: calcareous clay; marl; lime mud; shell layers; sand, silts. Carbonate minerals: aragonite; calcite (Lo-Mg); calcite (Hi-Mg); dolomite. Ooliths: Trace; common; abundant. Numbers left indicate depth (cm).

Much of interest can also be gained from a study of former lake levels—not only for palaeolimnologists but for climatologists, geologists and archaeologists. Indications of former low levels include the occurrence of coarse sediments offshore, missing pollen zones (see below for further explanation), and changing patterns of distribution of littoral aquatic plants. Abandoned beach terraces are particularly significant indications of former high lake levels. The occurrence of raised terraces has been especially important in demonstrating that Lake Corangamite in Victoria was formerly much more extensive than it is today.

Sedimentological studies of even dry lake beds and of lunettes (crescentic dunes associated with present day or former lakes) can also contribute much of palaeolimnological interest. Indeed, it has been said (Bowler 1971) that 'some of the best evidence of past hydrologic and geochemical environments is provided by the transverse and crescentic dunes or lunettes which occur on the eastern shores of most lake basins in the semi-arid zone of both southeastern and southwestern Australia'. The previous hydrology of the Willandra series of lakes in southwestern New South Wales (all now dry) was determined largely by analysis of sediment structure in samples from dry lake beds and associated lunettes.

A great deal more information is available in sediments than can be gained by examination of their physical structure alone. Sediments also contain much chemical evidence. Even the most simple of chemical analyses, the determination of the ratio of organic to inorganic matter, is frequently very instructive. Sediments rich in organic matter imply high biological productivities and eutrophic conditions at the time of their deposition, whereas those poor in organic matter imply low productivities and oligotrophic conditions.

At a more sophisticated level, interpretive use has been made of a large number of specific inorganic substances preserved in sediments. These include sodium, potassium, calcium, magnesium, manganese, iron, phosphorus, nitrogen, sulphates, (bi)carbonates, silicon, titanium, aluminium, gallium, zinc, copper and others. Only a few examples of their use can be given.

The ratio of manganese and iron in lake sediments can be informative about the nature of catchment soils, the extent to which lake hypolimnia became deoxygenated, and hence how productive lakes have been. The concentration of phosphorus in sediments can also tell much about former lake productivities. The recent large increases of phosphorus concentrations in the sediments of many eutrophied lakes provide the palaeolimnological record of the discovery of detergents! High inputs of silicon and bicarbonate (*vis-à-vis* low inputs of silicon and a dominance of chloride) into the sediments of some closed and saline lakes have been correlated with weathering rates and thus climate. The input of biologically inert elements such as titanium allows even better estimation of weathering rates than does the input of biologically reactive elements. Carbonate analysis provides a final example. Carbonate determinations for the sediments of some western Victorian lakes have been used to interpret former salinities (and thus palaeoclimates).

A variety of organic compounds has also been isolated from lake sediments: some of the more important are amino acids, carbohydrates, chlorophyllous, flavinoid and carotenoid pigments, and alkanes (saturated hydrocarbons). There are many more, e.g. phycobilins and porphyrins. However, there are many difficulties involved in their use as palaeolimnological indices. It is, for example, difficult to be certain as to whether the organic compound was produced *in situ* or in the catchment basin. Differential rates of degradation are another difficulty. Degradation itself is, of course, another. And it is not easy to distinguish the origin and synthetic mechanisms for many organic molecules. For these and other reasons, it appears that although a good deal of information is available in organic compounds in lake sediments, it is of limited palaeolimnological use at present. Nevertheless, the field promises much.

The one exception to the equivocatory value of organic evidence is provided by preserved plant pigments: so-called fossil pigments. Upon the death of a plant, pigments are degraded by loss of side groups and ions (e.g. Mg) and the products tend to decrease in solubility and increase in stability. Thus:

Chlorophyll *a* itself is rare, but the other degradation products are common, together with bacteriochlorophyll degradation products, and several carotenoids (some specific for certain groups, e.g. the cyanobacteria). Large pigment concentrations in recent sediments can clearly be correlated with rapid eutrophication of many lakes in recent decades and other anthropogenic phenomenon, e.g. deforestation. Sometimes, however, a decrease in pigment concentrations may signify a productivity decrease induced by man.

Though less useful than pigments, alkanes (saturated aliphatic hydrocarbons of chemical formula C_nH_{2n+2}) may also provide some clues to past history. Alkanes from C_{15} onwards are colourless solids and occur in plants. Higher plant residues yield mostly C_{27}, C_{29}, C_{31}, and C_{33} alkanes and small amounts only of C_{26} to C_{34} alkanes. In algae and other lower plants, residues are C_{26} to C_{34} with C_{17} the main alkane component. Thus, the relative contributions of higher plants and algae to lake sediments can be estimated. Estimates have even been made on sediments of Eocene age (in this case, the alkanes were clearly from algae, proving that the sediments were laid down in lake conditions).

BIOLOGICAL RECORDS

Many plant and animal groups leave remains which become preserved in lake sediments. Although only certain parts of an animal or plant are usually preserved, these remains can nevertheless frequently be related to particular species. When only parts of a lifecycle are preserved, then identification is usually possible only to genus or family.

Plant remains comprise pollen, diatoms, some green algae, Chrysophyceae, dinoflagellates, parts of blue-green algae, and various fractions of other plant tissues including charcoal. Of these, pollen is undoubtedly the most significant, and all comprehensive palaeolimnological studies involve an analysis of pollen stratigraphy. However, it is important to note that it is the vegetation of the *drainage basin* that is the principal subject of analysis. Local conditions in the lake are indicated by the pollen and spores of aquatic vegetation and by algal remains. Pollen stratigraphy, defined, is the record of the chronological and geographical distribution of fossil pollen and spores preserved in sediments (not necessarily those of lakes only). As such, pollen stratigraphy is a branch of palynology, the study of all aspects of pollen and spores. Pollen comprises microscopic (5–200 μm) reproductive structures produced by the male reproductive apparatus of the higher plants (angiosperms and gymnosperms). Morphologically complex, pollen grains have one to many apertures that are round to ellipsoidal (usually), or elongate, and have complex surface patterns enabling in most cases precise identification of source plants. Spores (5–200 μm) are reproductive bodies produced by the lower vascular and by non-vascular plants. Typically they are kidney-bean shaped or triangular, and have smooth or ornamented surfaces. Both pollen and spores have outer walls (the exine) which are highly resistant to decay; the compound involved has been termed sporopollenin (though apparently little is known of its chemical construction). The important point in relation to pollen stratigraphy is that changes in pollen spectra from one level to the next can be related to climatic and environmental changes in the drainage basin over periods of time. It is this dynamic aspect that gives pollen stratigraphy its importance.

It is only in the last decade or so that pollen stratigraphies for Australian lake sediments have been constructed. For the most part these relate to freshwater and relatively young coastal swamps, or crater lakes and swamps in regions of recent volcanic action (Atherton Tablelands of northeastern Queensland, western Victoria, southeastern South Australia). A few are concerned with older lakes such as Lake George near Canberra, and one with a clearly saline lake, Lake Frome (Fig. 13.2). The analysis of pollen in Lake Frome sediments deserves special mention for until recently most palynologists believed that salt lake sediments were likely to be poor sources of palaeolimnological information. It was no doubt a matter of great relief to Australian palaeolimnologists to find that not so.

Despite the increased availability of Australian pollen stratigraphy, their limited number and lack of geographical spread do not yet permit any detailed analysis of palaeoclimates during the Quaternary on a continent-wide basis. However, they do indicate that climatic instability and *overall* aridity appear to have been marked features of Australia for at least the Quaternary period (present to 1 million years ago).

Charcoal in lake sediments, though less useful than pollen as an indicator of past conditions, is of particular value in Australia. Its appearance in significant quantities in lake cores has been related to the presence of man; the assumption is that man increased the frequency of fires. The earliest known human remains in Australia are about 36 000 years old, but on the evidence provided by charcoal in lake sediments, man may have been in Australia for a good deal longer.

Of plant remains preserved in sediments and derived from lake floras *in situ* (as opposed to catchment floras), and therefore more directly indicative of palaeolacustrine conditions, diatoms are the most important. There are various interpretive problems, but diatom stratigraphy has much value since extrapolations backwards can be made from the known ecological requirements of present-day diatoms. Thus, as a very general rule, centric diatoms are associated with oligotrophic environments of low productivity, and pennate ones with eutrophic, productive conditions. Diatom stratigraphy may also have a useful role in ecological studies of modern diatoms. Studies of the diatoms in the sediments of some western Victorian lakes have been made, and whilst not extensive are of considerable interest. They are of special interest as correlates of other palaeolimnological evidence.

Though less important than diatom remains, pollen and spores from *in situ* aquatic macrophytes may also provide useful palaeolimnological evidence. They have been used successfully to provide such evidence for Lake George.

With regard to evidence provided by animal remains, it has been claimed that almost all groups of animals leave at least some fragments of their anatomy for subsequent preservation. However, the number and sorts of

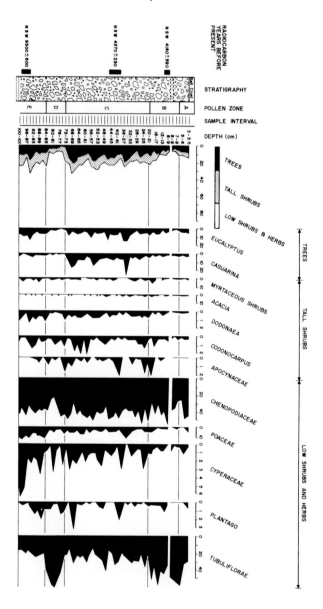

Fig. 13.2 Fossil pollen stratigraphy at Lake Frome, South Australia. The pollen data are recorded as percentages of total land pollen. After Singh (1981). ⠿ laminated alcy with small gypsum crystals; ≈ disconformity; ● <1%.

animal fragments usually preserved are relatively limited in scope. They comprise certain parts of Cladocera and midges (Chironomidae, Ceratopogonidae, and Chaoborinae) and ostracod valves; rhizopod tests are also preserved in special circumstances, as also are sponge, rotifer, turbellarian and bryozoan fragments. Larger preserved animal remains are mostly mollusc shells, beetle fragments and fish scales and vertebrae. Quite why there is such differential preservation is not always clear.

Cladoceran remains include the carapace, ephippial shell, head shield, postabdomen, mandible and a few other minor fragments. These usually enable identification to be taken to species level. Chydorid Cladocera, relative to other cladoceran families, are especially likely to be preserved since they have a thicker carapace than most Cladocera. No detailed studies of cladoceran remains in Australian lake sediments have yet been made. This undoubtedly reflects the rather poor state of our taxonomic knowledge of modern forms (see Table 4.1). Nevertheless, the occurrence of ephippia of the salt-tolerant species *Daphniopsis pusilla* in some cores from lakes in western Victoria has been noted and used as further evidence of saline conditions in these lakes.

There were also no detailed studies of ostracods in Australian lake sediments until recently. The firm taxonomic and ecological foundation now available for this group of crustaceans in Australia has, however, enabled significant advances to be made in the use of ostracods as palaeolimnological tools. Of particular significance here is the fact that in Australia ostracods have many species and these inhabit a wide variety of lake environments, from fresh to highly saline. To illustrate the sort of results now possible in ostracod stratigraphy in Australia, data from a sediment core of Lake Keilambete are given in Fig. 13.3. In this core, the presence of *M. praenuncia* indicates that the salinity was between 7 and 42°/∘∘, and, when co-occurring with *P. baueri*, between 20 and 43°/∘∘. *A. robusta* indicates a salinity between 7 and 145°/∘∘, but, when co-occurring with large numbers of *D. compacta*, a salinity of between 45 and 77.5°/∘∘. The presence of *L. lacustris* indicates permanent water with a salinity below 35°/∘∘.

The most common insect remains are the head capsules of midge larvae, particularly those of the family Chironomidae. This is fortunate, for a great deal of work has been directed towards relating living chironomid species with lake types: as a general rule in the northern hemisphere, oligotrophic lakes are characterized by *Tanytarsus*, mesotrophic lakes by *Stictochironomus* and *Sergentia*, and eutrophic lakes by *Chironomus*. This scheme does not seem to apply in Australia (see chapter 2). However, at least in the northern hemisphere, it would appear simple to relate chironomid remains with lake type at the time of deposition. In practice, it is not so simple, and there are

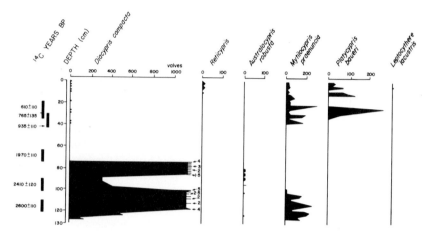

Fig. 13.3 Distribution of fossil ostracods in a core from Lake Keilambete, Victoria. Numbers in the graph for *Diacypris compacta* refer to the quantity of ostracod valves recovered per 3 g of sediment. Dots refer to low amounts of fossils. Redrawn after De Deckker (1981).

many interpretive difficulties. One Australian study of chironomid head capsules was in Lake Werowrap, a shallow saline lake in western Victoria. The differential occurrence of head capsules of three species (*Tanytarsus barbitarsis, Chironomus duplex* and *Procladius paludicola*) in a core from the lake was related to past changes in the salinity of the lake. When salinity was high *T. barbitarsis* occurred, sometimes with *P. paludicola*, a facultative predator; when salinity was low, only *C. duplex* occurred.

Ceratopogonid remains are rarer: all are of larvae, and again it is the head capsule that is preserved. Chaoborid remains mainly take the form of larval mandibles, though other fragments may be preserved (respiratory horns of pupae). As chaoborid larvae are said to characterize highly eutrophic lakes, the occurrence of their remains implies the occurrence of high rates of organic deposition and productivity, and low concentrations of oxygen at the time of preservation. Chaoborid larvae, however, have been found in quite nutrient-poor impoundments in Australia, so it is clear that all is not as simple as might be implied.

Much less need be said about other animal remains. However, shell remains of *Coxiella,* a snail of Australian salt lakes, have proved useful indicators of past high salinities in several lakes. Turbellarian remains comprise egg cocoons, bryozoan remains are statoblasts, poriferans are represented by their characteristic spicules, and rotifers by resting eggs.

References

1 INTRODUCTION

AUSTRALIAN WATER RESOURCES COUNCIL (1981) *The First National Survey of Water Use in Australia.* Australian Government Publishing Service, Canberra.

BAYLY I.A.E. (1979) Aquatic science research in Australia since 1950: an analysis of trends. *Australian Fisheries,* December 1979, 21–4.

BURTON J.R., WILLIAMS W.D., ST JOHN E. and HILL D. (1974) *The Flooding of Lake Pedder.* Lake Pedder Committee of Enquiry. Final Report, April 1974. Australian Government Publishing Service, Canberra.

COLE G.A. (1975) *Textbook of Limnology.* Mosby, St Louis.

DEPARTMENT OF NATIONAL RESOURCES (1976) *Review of Australia's Water Resources 1975.* Australian Government Publishing Service, Canberra.

GOLDMAN C.R. and HORNE A.J. (1983) *Limnology.* McGraw-Hill, New York.

GOLTERMAN H.L. (1975) *Physiological Limnology: An Approach to the Physiology of Lake Ecosystems.* Elsevier, Amsterdam.

HUTCHINSON G.E. (1957) *A Treatise in Limnology. Volume I – Geography, Physics, and Chemistry.* Wiley, New York.

HUTCHINSON G.E. (1967) *A Treatise in Limnology. Volume II – Introduction to Lake Biology and the Limnoplankton.* Wiley, New York.

HUTCHINSON G.E. (1975) *A Treatise in Limnology. Volume III – Limnological Botany.* Wiley, New York.

JACOBI R. (1981) Institute of Freshwater Studies Bill 1981. *Hansard,* 4 June 1981, House of Representatives, Canberra, 3135–3141.

MOSS B. (1980) *Ecology of Fresh Waters.* Blackwell Scientific Publications, Oxford.

MUNRO C.H. (1974) *Australian Water Resources and their Development.* Angus and Robertson, Sydney.

O'LOUGHLIN E. M. and CULLEN, P. (eds) (1982) *Prediction in Water Quality.* Australian Academy of Science, Canberra.

SENATE STANDING COMMITTEE ON NATIONAL RESOURCES (1978) *The Commonwealth's Role in the Assessment, Planning, Development and Management of Australia's Water Resources.* Australian Government Publishing Service, Canberra.

SERVENTY V. and RAYMOND R. (1980) *Lakes and Rivers of Australia.* Summit Books, Sydney.

WETZEL R.G. (1975) *Limnology.* Saunders, Philadelphia.

WILLIAMS W.D. (ed.) (1980) *An Ecological Basis for Water Resource Management.* Australian National University Press, Canberra.

2 LAKES AND OTHER STANDING WATERS

BAYLY I.A.E. (1967) The general biological classification of aquatic environments with special reference to those in Australia. In A.H. Weatherley (ed.) *Australian Inland*

Waters and their Fauna: Eleven Studies. Australian National University Press, Canberra.

BAYLY I.A.E. (1972) Salinity tolerance and osmotic behaviour of animals in athalassic saline and marine hypersaline waters. *Ann. Rev. Ecol. Systematics* **3**, 233–68.

BAYLY I.A.E. and MORTON D.W. (1978) Aspects of the zoogeography of Australian microcrustaceans. *Verh. Int. Ver. Limnol.* **20**, 2537–40.

BAYLY I.A.E. and WILLIAMS W.D. (1966) Chemical and biological studies on some saline lakes of south-east Australia. *Aust. J. Mar. Freshwat. Res.* **17**, 177–228.

BAYLY I.A.E. and WILLIAMS W.D. (1973) *Inland Waters and their Ecology.* Longman, Melbourne.

BISHOP J.A. (1967) Seasonal occurrence of a branchiopod crustacean, *Limnadia stanleyana* King (Conchostraca) in eastern Australia. *J. Anim. Ecol.* **36**, 77–95.

BRIGGS S.V. (1981) Freshwater wetlands. In R.H. Groves (ed.) *Australian Vegetation.* Cambridge University Press, Cambridge.

BROCK M.A. (1981) Accumulation of proline in a submerged aquatic halophyte, *Ruppia* L. *Oecologia,* **51**, 217–19.

DE DECKKER P. and GEDDES M.C. (1980) Seasonal fauna of ephemeral saline lakes near the Coorong Lagoon, South Australia. *Aust. J. Mar. Freshwat. Res.* **31**, 677–99.

DEW B. (1963) Animal life in caves. *Aust. Mus. Mag.* **15**, 158–61.

DYCE A.L. (1964) 'Tree holes' in our drier inland areas. *Aust. Soc. Limnol. Newsl.* **3**(1), 8–9.

EDWARD D.H. (1964) A cryptobiotic chironomid from south-western Australia. *Aust. Soc. Limnol. Newsl.* **6**(1), 29–30.

EDWARD D.H. (1968) Chironomidae in temporary freshwaters. *Aust. Soc. Limnol. Newsl.* **6**(1), 3–5.

FRITH H.J. (1959) The ecology of wild ducks in inland New South Wales. 1. Waterfowl habitats. *CSIRO Wildlife Res.* **4**, 1–97.

GEDDES M.C. (1976) Seasonal fauna of some ephemeral saline waters in western Victoria with particular reference to *Parartemia zietziana* Sayce (Crustacea: Anostraca). *Aust. J. Mar. Freshwat. Res.* **27**, 1–22.

GEDDES M.C., DE DECKKER P., WILLIAMS W.D., MORTON D.W. and TOPPING M. (1981) On the chemistry and biota of some saline lakes in Western Australia. In W.D. Williams (ed.) *Salt Lakes.* Junk, The Hague.

GOOD, R.E., WHIGHAM D.F. and SIMPSON R.L. (eds) (1978) *Freshwater Wetlands. Ecological Processes and Management Potential.* Academic Press, New York.

GRANT J.W.G. and BAYLY I.A.E. (1981) Predator induction of crests in morphs of the *Daphnia carinata* King complex. *Limnol. Oceanogr.* **26**, 201–18.

HAMMER U.T. (1981) A comparative study of primary production and related factors in four saline lakes in Victoria, Australia. *Int. Rev. Ges. Hydrobiol.* **66**, 701–43.

HEBERT P.D. (1977) A revision of the taxonomy of the genus *Daphnia* (Crustacea; Daphnidae) in south-eastern Australia. *Aust. J. Zool.* **25**, 371–98.

HEBERT P.D. (1978) Cyclomorphosis in natural populations of *Daphnia carinata* King. *Freshwater Biol.* **8**, 79–90.

JONES R.E. (1974) The effects of size-selective predation and environmental variation on the distribution and abundance of a chironomid, *Paraborniella tonnoiri* Freeman. *Aust. J. Zool.* **22**, 71–89.

KIRK J.T.O. (1976) Yellow substance (Gelbstoff) and its contribution to the attenuation of photosynthetically active radiation in some inland and coastal south-eastern Australian waters. *Aust. J. Mar. Freshwat. Res.* **27**, 61–71.

KIRK J.T.O. (1977) Attenuation of light in natural waters. *Aust. J. Mar. Freshwat. Res.* **28**, 497–508.

KIRK J.T.O. (1980) Spectral absorption properties of natural waters: contributions of the soluble and particulate fractions to light absorption in some inland waters of south-eastern Australia. *Aust. J. Mar. Freshwat. Res.* **31**, 287–96.

LAKE P.S. (1977) Pholeteros — the faunal assemblage found in crayfish burrows. *Aust. Soc. Limnol. Newsl.* **15**(1), 57–60.

MITCHELL B.D. (1978) Cyclomorphosis in *Daphnia carinata* King (Crustacea: Cladocera) from two adjacent sewage lagoons in South Australia. *Aust. J. Mar. Freshwat. Res.* **29**, 565–76.

MITCHELL B.D. (in press) Limnology of mound springs and temporary pools south and west of Lake Eyre. Nature Conservation Society of South Australia, Adelaide.

MITCHELL B.D. (1980) Waste stabilization ponds. In W.D. Williams (ed.) *An Ecological Basis for Water Resource Management.* Australian National University Press, Canberra.

MORRISSY N.M. (1974) The ecology of marron *Cherax tenuimanus* (Smith) introduced into some farm dams near Boscabel in the Great Southern Area of the Wheat Belt Region of Western Australia. *Fish. Res. Bull. West. Aust.* **12**, 1–55.

MORTON D.W. and BAYLY I.A.E. (1977) Studies on the ecology of some temporary freshwater pools in Victoria with special reference to microcrustaceans. *Aust. J. Mar. Freshwat. Res.* **28**, 439–54.

SHIEL R.J. (1976) Associations of Entomostraca with weedbed habitats in a billabong of the Goulburn River, Victoria. *Aust. J. Mar. Freshwat. Res.* **27**, 533–49.

SHIEL R.J. (1980) Billabongs of the Murray-Darling system. In W.D. Williams (ed.) *An Ecological Basis for Water Resource Management.* Australian National University Press, Canberra.

TIMMS B.V. (1980) The benthos of Australian lakes. In W.D. Williams (ed.) *An Ecological Basis for Water Resource Management.* Australian National University Press, Canberra.

TIMMS B.V. (1980) Farm dams. In W.D. Williams (ed.) *An Ecological Basis for Water Resource Management.* Australian National University Press, Canberra.

TIMMS B.V. (1981) Animal communities in three Victorian lakes of differing salinity. In W.D. Williams (ed.) *Salt Lakes.* Junk, The Hague.

TIMMS B.V. (1982) Coastal dune waterbodies of north-eastern New South Wales. *Aust. J. Mar. Freshwat. Res.* **33**, 203–22.

WALKER K.F. (1973) Studies on a saline lake ecosystem. *Aust. J. Mar. Freshwat. Res.* **24**, 21–7.

WEATHERLEY A.H. (1967) The inland waters of Australia: Introductory. In A.H. Weatherley (ed.) *Australian Inland Waters and their Fauna: Eleven Studies.* Australian National University Press, Canberra.

WILLIAMS W.D. (1967) The chemical characteristics of lentic surface waters in Australia: A review. In A.H. Weatherley (ed.) *Australian Inland Waters and their Fauna: Eleven Studies.* Australian National University Press, Canberra.

WILLIAMS W.D. (1975) A note on the macrofauna of a temporary rainpool in semi-arid Western Australia. *Aust. J. Mar. Freshwat. Res.* **26**, 425–9.

WILLIAMS W.D. (1980) Distinctive features of Australian water resources. In W.D. Williams (ed.) *An Ecological Basis for Water Resource Management.* Australian National University Press, Canberra.

WILLIAMS W.D. (1981) Athalassic (inland) salt lakes: an introduction. In W.D. Williams (ed.) *Salt Lakes.* Junk, The Hague.

WILLIAMS W.D. (1981) The limnology of saline lakes in Victoria: A review of some recent studies. In W.D. Williams (ed.) *Salt Lakes*. Junk, The Hague.

WILLIAMS W.D. and BUCKNEY R.T. (1976) Stability of ionic proportions in five salt lakes in Victoria, Australia. *Aust. J. Mar. Freshwat. Res.* **27**, 367–77.

3   RIVERS AND STREAMS

BAILEY P.C.E. (1981) Insect drift in Condor Creek, Australian Capital Territory. *Aust. J. Mar. Freshwat. Res.* **32**, 111–20.

BAYLY I.A.E. and WILLIAMS W.D. (1973) *Inland Waters and Their Ecology*. Longman, Melbourne.

BISHOP K.A., ALLEN S.A., POLLARD D.A. and COOK M.G. (1981) *Ecological studies on freshwater fish of the Alligator Rivers region, Northern Territory*. Report to the Office of the Supervising Scientist, New South Wales Fisheries, Sydney.

BLACKBURN W.M. and PETR T. (1979) Forest litter decomposition and benthos in a mountain stream in Victoria, Australia. *Arch. Hydrobiol.* **86**, 453–98.

BUCKNEY R.T. (1977) Chemical dynamics in a Tasmanian river. *Aust. J. Mar. Freshwat. Res.* **28**, 261–8.

BUCKNEY R.T. (1979) Chemical loadings in a small river, with observations on the role of suspended matter in the nutrient flux. *AWRC Tech. Pap.* No. 40.

CADWALLADER P.L. and EDEN A.K. (1977) Effect of a total solar eclipse on invertebrate drift in Snob's Creek, Victoria. *Aust. J. Mar. Freshwat. Res.* **28**, 799–805.

CAMPBELL I.C. (1980) Diurnal variations in the activity of *Mirawara purpurea* Riek (Ephemeroptera, Siphlonuridae) in the Aberfeldy River, Victoria, Australia. In J.F. Flannagan and K.E. Marshall (eds) *Advances in Ephemeroptera Biology*. Plenum, New York.

CHANEY J.A., THOMAS D.P. and TYLER P.A. (1979) *Diatoms and other freshwater algae of the Magela Creek system as monitors of heavy metal pollution*. First Interim Report to the Office of the Supervising Scientist; Botany Department, University of Tasmania.

COLBO M.H. and MOORHOUSE D.E. (1979) The ecology of pre-imaginal Simuliidae (Diptera) in south-east Queensland, Australia. *Hydrobiologia* **63**, 63–79.

CUMMINS K.W. (1979) The natural stream ecosystem. In J.V. Ward and J.A. Stanford (eds) *The Ecology of Regulated Streams*. Plenum, New York.

CUMMINS K.W. and KLUG M.J. (1979) Feeding ecology of stream invertebrates. *Ann. Rev. Ecol. Syst.* **10**, 147–72.

DEPARTMENT OF NATIONAL RESOURCES, AUSTRALIAN WATER RESOURCES COUNCIL (1976) *Review of Australia's Water Resources 1975*. Australian Government Publishing Service, Canberra.

DEPARTMENT OF NATIONAL RESOURCES, AUSTRALIAN WATER RESOURCES COUNCIL (1978) *Variability of Runoff in Australia*. AWRC Hydrological Series, No. 11. Australian Government Publishing Service, Canberra.

DOEG T.J. and LAKE P.S. (1981) A technique for assessing the composition and density of the macroinvertebrate fauna of large stones in streams. *Hydrobiologia,* **80**, 3–6.

FRITH H.J. and SAWER G. (eds) (1974) *The Murray Waters. Man, Nature and a River System*. Angus and Robertson, Sydney.

GLOVER C.J.M. and SIM T.C. (1978) Studies on central Australian fishes: a progress report. *South Aust. Nat.* **52**(3), 35–44.

HART B.T. and McGREGOR R.J. (1980) Limnological survey of eight billabongs in the Magela Creek catchment, Northern Territory. *Aust. J. Mar. Freshwat. Res.* **31**, 611–26.

HAY D.A. and BALL I.R. (1979) Contributions to the biology of freshwater planarians (Turbellaria) from the Victorian Alps, Australia. *Hydrobiologia,* **62**, 137–64.

HYNES H.B.N. (1970) *The Ecology of Running Waters.* Liverpool University Press, Liverpool.

HYNES H.B.N. (1975) The stream and its valley. *Verh. Int. Verein. Limnol.* **19**, 1–15.

HYNES H.B.N. and HYNES M.E. (1975) The life histories of many of the stoneflies (Plecoptera) of south-eastern mainland Australia. *Aust. J. Mar. Freshwat. Res.* **26**, 113–53.

KENDRICK G.W. (1976) The Avon: faunal and other notes on a dying river in south-western Australia. *Western Australian Naturalist* **13**(5), 97–114.

KOSTE W. (1981) Zur Morphologie, Systematik und Ökologie von neuen monogononten Rädertieren (Rotatoria) aus dem Überschwemmungsgebiet des Magela Creek in der Alligator-River-Region Australiens, N.T. Teil 1. *Osnabrücker naturwiss. Mitt.* **8**, 97–126.

KOSTE W. and SHIEL R.J. (in press) On the morphology, systematics and ecology of new monogonont Rotifera (Rotatoria) from the Magela Creek, Alligator River region N.T., Australia, II. *Trans. R. Soc. S. Aust.* **107.**

LAKE J.S. (1967) Principal fishes of the Murray-Darling River system. In A.H. Weatherly (ed.) *Australian Inland Waters and their Fauna: Eleven Studies.* Australian National University Press, Canberra.

LAKE P.S. (1981) *Ecology of the macroinvertebrates of Australian upland streams – a review of current knowledge.* Paper presented at the Annual Congress of the Australian Society for Limnology, May 1981.

MARCHANT R. (1982) Life spans of two species of tropical mayfly nymph (Ephemeroptera) from Magela Creek, Northern Territory. *Aust. J. Mar. Freshwat. Res.* **33**, 173–9.

MARCHANT R. (1982) Seasonal variation in the macroinvertebrate fauna of billabongs along Magela Creek, Northern Territory. *Aust. J. Mar. Freshwat. Res.* **33**, 329–42.

MORRISSY N.M. (1974) Reversed longitudinal salinity profile of a major river in the south-west of Western Australia. *Aust. J. Mar. Freshwat. Res.* **25**, 327–35.

MORRISSY N.M. (1979) Inland (non-estuarine) halocline formation in a Western Australian river. *Aust. J. Mar. Freshwat. Res.* **30**, 343–53.

MUIR G.L. and JOHNSON W.D. (1978) Chemistry of the Bogan River, New South Wales, with special reference to the sources of dissolved material. *Aust. J. Mar. Freshwat. Res.* **29**, 399–407.

PIDGEON R.W.J. and CAIRNS S.C. (1981) Decomposition and colonization by invertebrates of native and exotic leaf material in a small stream in New England (Australia). *Hydrobiologia* **77**, 113–27.

POTTER I.C., CANNON D. and MOORE J.W. (1975) The ecology of algae in the Moruya River, Australia. *Hydrobiologia* **47**, 415–30.

SHIEL R.J. (1978) Associations of entomostraca with weedbed habitats in a billabong of the Goulburn River, Victoria. *Aust. J. Mar. Freshwat. Res.* **27**, 533–49.

SHIEL R.J. (1978) Zooplankton communities of the Murray-Darling system – a preliminary report. *Proc. Roy. Soc. Vict.* **90**(1), 193–202.

SHIEL R.J. (1979) Synecology of the Rotifera of the River Murray, South Australia. *Aust. J. Mar. Freshwat. Res.* **30**, 255–63.

SHIEL R.J. (1981) *Plankton of the Murray-Darling river system, with particular reference to the zooplankton.* Ph.D. thesis, University of Adelaide.

SHIEL R.J., WALKER K.F. and WILLIAMS W.D. (1982) Plankton of the lower River Murray. *Aust. J. Mar. Freshwat. Res.* **33**, 301–27.

SMITH M.J. and WILLIAMS W.D. (1982) A taxonomic revision of Australian species of *Atyoida* Randall (Crustacea, Decapoda, Atyidae), with remarks on the taxonomy of the genera *Atyoida* and *Atya* Leach. *Aust. J. Mar. Freshwat. Res.* **33**, 343–61.

SMITH M.J. and WILLIAMS W.D. (in press) Reproductive strategies of some freshwater amphipods in southern Australia. *Rec. Aust. Mus.*

SUTER P.J. and WILLIAMS W.D. (1977) Effects of a total solar eclipse on stream drift. *Aust. J. Mar. Freshwat. Res.* **28**, 793–8.

TAIT R.D. (1981) Comparison of the diets of the northern spotted barramundi (*Scleropages jardini*) and the giant perch (*Lates calcarifer*) in northern Australia. *Verh. Int. Verein. Limnol.* **21**, 1320–5.

TOWNS D.R. (in press) Terrestrial oviposition by two species of caddisfly in South Australia (Trichoptera: Leptoceridae). *J. Aust. Ent. Soc.*

TOWNSEND C.R. (1980) *The Ecology of Streams and Rivers.* Edward Arnold, London.

WALKER K.F. (1979) Regulated streams in Australia: The Murray-Darling system. In J.V. Ward and J.A. Stanford (eds) *The Ecology of Regulated Streams.* Plenum, New York.

WALKER K.F. (1981). Ecology of freshwater mussels in the River Murray. *AWRC Tech. Paper* No. 63.

WALKER K.F. and HILLMAN T.J. (1977) *Limnological Survey of the River Murray in Relation to Albury/Wodonga, 1973/76.* Albury-Wodonga Development Corporation and Gutteridge, Haskins and Davey, Melbourne.

WALKER K.F., HILLMAN T.J. and WILLIAMS W.D. (1978) The effects of impoundments on rivers: An Australian case study. *Verh. Int. Verein. Limnol.* **20**, 1695–1701.

WALKER T.D. and TYLER P.A. (1979) *A limnological survey of the Magela Creek System, Alligator Rivers region, Northern Territory.* First and Second Interim Report to the Office of the Supervising Scientist; Botany Department, University of Tasmania, Hobart.

WHITTON B.A. (ed.) (1975) *River Ecology.* Blackwell Scientific Publications, Oxford.

WIGGINS G.B., MACKAY R.J. and SMITH I.M. (1980) Evolutionary and ecological strategies of animals in annual temporary pools. *Arch. Hydrobiol. Suppl.* **58**, 97–206.

WILLIAMS D.D. and HYNES H.B.N. (1977) The ecology of temporary streams. II. General remarks on temporary streams. *Int. Rev. Ges. Hydrobiol.* **62**, 53–61.

WILLIAMS W.D. (1977) Some aspects of the ecology of *Paratya australiensis* (Crustacea, Decapoda, Atyidae) in Australia. *Aust. J. Mar. Freshwat. Res.* **28**, 403–15.

WILLIAMS W.D. (1982) Running water ecology in Australia. In M. Lock and D.D. Williams (eds) *Perspectives in Running Water Ecology.* Plenum, New York.

4   INVERTEBRATES

ANDERSON D.T., FLETCHER M.J. and LAWSON-KERR C. (1976) A marine caddis fly, *Philanisus plebeius*, ovipositing in a starfish, *Patriella exigua. Search* **7** (11–12), 483–4.

ANDERSON D.T. and LAWSON-KERR C. (1977) The embryonic development of the marine caddis fly, *Philanisus plebeius* Walker (Trichoptera: Chathamidae). *Biol. Bull.* **153**, 98–105.

ASHBURNER L.D. (1876) Fish diseases and potential fish diseases in Australia. *Animal Quarantine (Commonwealth Department of Health)* **5**(1), 1–7.

BAYLY I.A.E. and ELLIS P. (1969) *Haloniscus searlei* Chilton: an aquatic 'terrestrial' isopod with remarkable powers of osmotic regulation. *Comp. Biochem. Physiol.* **31**, 523–8.

BERG C.O. and KNUTSON L. (1978) Biology and systematics of the Sciomyzidae. *Ann. Rev. Entomol.* **23**, 239–58.

BLACK R.H. (1972) *Malaria in Australia.* Australian Government Publishing Service, Canberra.

BLAIR D. and FINLAYSON C.M. (1981) Observations on the habitat and biology of a lymnaeid snail *Austropeplea vinosa* (Gastropoda: Pulmonata), an intermediate host for avian schistosomes in tropical Australia. *Aust. J. Mar. Freshwat. Res.* **32**, 757–67.

BORAY J.C. (1964) Studies on the ecology of *Lymnaea tomentosa*, the intermediate host of *Fasciola hepatica*. 1. History, geographical distribution, and environment. *Aust. J. Zool.* **12**, 217–30.

CALMAN W.T. (1896) On the genus *Anaspides* and its affinities with certain fossil Crustacea. *Trans. Roy. Soc. Edinb.* **38**(4), 787–802.

CAMPBELL I.C. (1981) Biogeography of some rheophilous aquatic insects in the Australian region. *Aquatic Insects* **3**(1), 33–43.

CAMPBELL I.C. (1981) Biology, taxonomy and water quality monitoring in Australian streams. *Water* **8**(4), 11–3.

DOHERTY R.L. (1977) Arthropod-borne viruses in Australia. *Aust. J. Exp. Biol. Med. Sci.* **55**, 103–30.

ELLIS P. and WILLIAMS W.D. (1970) The biology of *Haloniscus searlei* Chilton, an oniscoid isopod living in Australian salt lakes. *Aust. J. Mar. Freshwat. Res.* **21**, 51–69.

FORSYTH J.R.L. (1980) Public health aspects of water usage. In W.D. Williams (ed.) *An Ecological Basis for Water Resource Management.* Australian National University Press, Canberra.

HARRISON L. (1928) On the genus *Stratiodrilus* (Archiannelida: Histriobdellidae), with a description of a new species from Madagascar. *Rec. Aust. Mus.* **16**, 116–21.

HASWELL W.A. (1893) A monograph on the Temnocephaleae. *Macleay Memorial Volume, Linn. Soc. N.S.W.* 93–152.

HASWELL W.A. (1900) On a new Histriobdellid. *Quart. J. Micr. Sci.* **43**, 299–335.

HICKMAN V.V. (1967) Tasmanian Temnocephalidea. *Pap. Proc. R. Soc. Tasm.* **101**, 227–50.

HUNTER D.M. and MOORHOUSE D.E. (1976) Comparative bionomics of adult *Austrosimulium pestilens* MacKerras and Mackerras and *A. bancrofti* (Taylor) (Diptera: Simuliidae). *Bull. Ent. Res.* **66**, 453–67.

HYNES H.B.N. (1976) Biology of Plecoptera. *Ann. Rev. Ent.* **21**, 135–53.

HYNES H.B.N. (1978) Annotated key to the stonefly nymphs (Plecoptera) of Victoria. *Aust. Soc. Limnol. Spec. Pub.* **2**, 1–63.

HYNES H.B.N. and HYNES M.E. (1975) The life histories of many of the stoneflies (Plecoptera) of south-eastern mainland Australia. *Aust. J. Mar. Freshwat. Res.* **26**, 113–53.

HYNES H.B.N. and HYNES M.E. (1980) The endemism of Tasmanian stoneflies (Plecoptera). *Aquatic Insects* **2**(2), 81–9.

JAMIESON J.A. (1975) *Studies of Amoebae of the Genus* Naegleria. M.Sc. Thesis, University of Adelaide.

JONES E.L. (1968) *Chironomus tepperi* Skuse (Diptera: Chironomidae) as a pest of rice in New South Wales. *Aust. J. Sci.* **31**, 89.

JONES W.G. and WALKER K.F. (1979) Accumulation of iron, manganese, zinc and cadmium by the Australian freshwater mussel *Velesunio ambiguus* (Phillipi) and its potential as a biological monitor. *Aust. J. Mar. Freshwat. Res.* **30**, 741–51.

JONES W.G. and WALKER K.F. (1979) An outline of biological monitoring in aquatic environments. *Water* **6**(2), 17–19.

KETTLE D.S. (1977) Biology and bionomics of bloodsucking ceratopogonids. *Ann. Rev. Entomol.* **22**, 33–51.

KETTLE, D.S., REYE E.J. and EDWARDS P.B. (1979) Distribution of *Culicoides molestus* (Skuse) (Diptera: Ceratopogonidae) in man-made canals in south-eastern Queensland. *Aust. J. Mar. Freshwat. Res.* **30**, 653–60.

KNOTT B. and LAKE P.S. (1980) *Eucrenonaspides oinotheke* gen. et sp.n. (Psammaspididae) from Tasmania, and a new taxonomic scheme for Anaspidacea (Crustacea, Syncarida). *Zoologica Scripta* **9**, 25–33.

LEADER J.P. (1972) Osmoregulation in the larva of the marine caddis fly, *Philanisus plebeius* (Walk). (Trichoptera). *J. Exp. Biol.* **57**, 821–38.

McMICHAEL D. and HISCOCK I.D. (1958) A monograph of the freshwater mussels (Mollusca: Pelecypoda) of the Australian region. *Aust. J. Mar. Freshwat. Res.* **9**, 372–507.

MILLS B.J., LAKE P.S. and SWAIN R. (1979) Two freshwater crustaceans suitable for toxicity tests in Australian waters. *Caulfield Institute of Technology, Water Studies Centre, Tech. Rept* No. 10.

MORRISSY N.M. (1980) Aquaculture. In W.D. Williams (ed.) *An Ecological Basis for Water Resource Management.* Australian National University Press, Canberra.

PETERS B.C. (1975) The control of the midge-fly (Diptera: Chironomidae) at Bolivar sewage treatment works. *E.W.S. Report 1228/68.* Adelaide. Cyclostyled.

PETR T. (1980) Medically important diseases with aquatic vectors and hosts. In W.D. Williams (ed.) *An Ecological Basis for Water Resource Management.* Australian National University Press, Canberra.

PILGRIM R.L.C. (1972) The aquatic larva and the pupa of *Choristella philpotti* Tillyard, 1917. *Pacific Insects* **14**(1), 151–68.

RIEK E.F. (1970) Mecoptera (Scorpion-flies). In CSIRO (sponsor) *Insects of Australia.* Melbourne University Press, Melbourne.

RIEK E.F. (1976) The marine caddisfly family Chathamiidae (Trichoptera). *J. Aust. Ent. Soc.* **1976**, 405–19.

RODHE K. (1978) The bird schistosome *Gigantobilharzia* sp. in the silver gull, *Larus novaehollandiae*, a potential agent of schistosome dermatitis in Australia. *Search* **9**, 40–2.

SOHN I.G. and KORNICKER L.S. (1972) Predation of schistosomiasis vector snails by Ostracoda (Crustacea). *Science, N.Y.* **175**, 1258–9.

STANLEY N.F. and ALPERS M.P. (eds) (1975) *Man-made Lakes and Human Health.* Academic Press, London.

THOMSON G.M. (1893) Notes on Tasmanian Crustacea, with descriptions of new species. *Proc. Roy. Soc. Tasm.* **189**, 45–76.

WALKER K.F. (1981) The distribution of freshwater mussels (Mollusca: Pelecypoda) in the Australian zoogeographic region. In A. Keast (ed.) *Ecological Biogeography in Australia*. Junk, The Hague.

WALKER K.F. (1981) Ecology of freshwater mussels in the River Murray. *A WRC Tech. Pap.* No. 63.

WILLIAMS J.B. (1980) Morphology of a species of *Temnocephala* (Platyhelminthes) ectocommensal on the isopod *Phreatoicopsis terricola*. *J. Nat. Hist.* **14**(2), 183–99.

WILLIAMS W.D. (1965) Ecological notes on Tasmanian Syncarida (Crustacea: Malacostraca), with a description of a new species of *Anaspides*. *Int. Rev. Ges. Hydrobiol.* **50**(1), 95–126.

WILLIAMS W.D. (1980) *Australian Freshwater Life: The Invertebrates of Australian Inland Waters*. 2nd edn. Macmillan, Melbourne.

WILLIAMS W.D. (1980) Biological monitoring. In W.D. Williams (ed.) *An Ecological Basis for Water Resource Management*. Australian National University Press, Canberra.

WILLIAMS W.D. (1981) The Crustacea of Australian inland waters. In A. Keast (ed.) *Ecological Biogeography of Australia*. Junk, The Hague.

WILLIAMS W.D. (1981) Aquatic insects: an overview. In A. Keast (ed.) *Ecological Biogeography of Australia*. Junk, The Hague.

WINTERBOURN M.J. (1980) The freshwater insects of Australasia and their affinities. *Palaeogeog. Palaeoclim. Palaeoecol.* **31**, 235–49.

ZWICK P. (1979) Revision of the stonefly family Eustheniidae (Plecoptera), with emphasis on the fauna of the Australian region. *Aquatic Insects* **1**(1), 17–50.

ZWICK P. (1981) Plecoptera. In A. Keast (ed.) *Ecological Biogeography of Australia*. Junk, The Hague.

5 FISH

ALLEN G. R. (1982) *Inland Fishes of Western Australia*. Western Australian Museum, Perth.

ANDREWS A.P. (1976) A revision of the family Galaxiidae (Pisces) in Tasmania. *Aust. J. Mar. Freshwat. Res.* **27**, 297–349.

ASHBURNER L.D. (1976) Australian fish face threat from foreign diseases. *Aust. Fish.* **35** (June), 18–22.

BELL J.D., BERRA T.M., JACKSON P.D., LAST P.R. and SLOANE R.D. (1980) Recent records of the Australian grayling *Prototroctes maraena* Günther (Pisces: Prototroctidae) with notes on its distribution. *Aust. Zool.* **20**(3), 419–31.

BERRA T.M. (1974) The trout cod, *Maccullochella macquariensis*, a rare freshwater fish of eastern Australia. *Biol. Conserv.* **6**, 53–6.

BERRA T. (1982) The life history of the Australian grayling, *Prototroctes maraena* Günther (Salmoniformes: Prototroctidae) in the Tambo River of Victoria. *Copeia* **1982**(4), 795–805.

BERRA T.M., MOORE R. and REYNOLDS L.F. (1975) The freshwater fishes of the Laloki River system of New Guinea. *Copeia* **1975**(3), 316–26.

BERRA T.M. and WEATHERLEY A.H. (1972) A systematic study of the Australian freshwater serranid fish genus *Maccullochella*. *Copeia* **1972**(1), 53–64.

BEUMER J.P. (1979) Feeding and movement of *Anguilla australis* and *A. reinhardtii* in Macleods Morass, Victoria, Australia. *J. Fish. Biol.* **14**, 573–92.

BISHOP K.A. and BELL, J.D. (1978) Observations on the fish fauna below Tallowa Dam (Shoalhaven River, New South Wales) during river flow stoppages. *Aust. J. Mar. Freshwat. Res.* **29**, 543-9.

BISHOP K.A. and BELL J.D. (1978) Aspects of the biology of the Australian grayling *Prototroctes maraena* Günther (Pisces: Prototroctidae). *Aust. J. Mar. Freshwat. Res.* **29**, 743-61.

CADWALLADER P.L. (1978) Some causes of the decline in range and abundance of native fish in the Murray-Darling River system. *Proc. Roy. Soc. Vict.* **90**, 211-24.

CADWALLADER P.L. (1979) Distribution of native and introduced fish in the Seven Creeks River system, Victoria. *Aust. J. Ecol.* **4**, 361-85.

CADWALLADER P.L., BACKHOUSE G.N. and FALLU R. (1980) Occurrence of exotic tropical fish in the cooling pondage of a power station in temperate south-eastern Australia. *Aust. J. Mar. Freshwat. Res.* **31**, 541-6.

CADWALLADER P.L., BACKHOUSE G.N., GOOLEY G.J. and TURNER J.A. (1979) New techniques for breeding and raising Murray Cod. *Aust. Fisheries* **38**(9).

CADWALLADER P.L. and ROGAN P.L. (1977) The Macquarie perch, *Macquaria australasica* (Pisces: Percichthyidae), of Lake Eildon, Victoria. *Aust. J. Ecol.* **2**, 409-18.

CHESSMAN B.C. and WILLIAMS W.D. (1974) Distribution of fish in inland saline waters in Victoria, Australia. *Aust. J. Mar. Freshwat. Res.* **25**, 167-72.

CONROY D.A. (1975) An evaluation of the present state of world trade in ornamental fish. *FAO Fisheries Technical Paper* **146**, 1-28.

COURTENEY W.R. JR and ROBINS C.R. (1973) Exotic aquatic organisms in Florida with emphasis on fishes: a review and recommendations. *Trans. Amer. Fish. Soc.* **102**, 1-12.

DAVIS T.L.O. (1977) Reproductive biology of the freshwater catfish, *Tandanus tandanus* Mitchell, in the Gwydir River, Australia. II. Gonadal cycle and fecundity. *Aust. J. Mar. Freshwat. Res.* **28**, 159-69.

DUNSTAN D.J. (1959) The barramundi *Lates calcarifer* (Bloch) in Queensland waters. *CSIRO Div. Fish. Oceanogr. Tech. Pap.* **5**.

FRANKENBERG R.S. (1966) Fishes of the family Galaxiidae. *Aust. Nat. Hist.* **15**, 161-4.

FRANKENBERG R.S. (1974) Native freshwater fish. In W.D. Williams (ed.) *Biogeography and Ecology in Tasmania.* Junk, The Hague.

GEDDES M.C. (1979) Salinity tolerance and osmotic behaviour of European carp (*Cyprinus carpio* L.) from the River Murray, Australia. *Trans. R. Soc. S. Aust.* **103**(7), 185-9.

GLOVER C.J.M. and SIM T.C. (1978) A survey of central Australian ichthyology. *Aust. Zool.* **19**(3), 245-56.

GLOVER C.J.M. and SIM T.C. (1978) Studies on central Australian fishes: a progress report. *South Aust. Nat.* **52**(3), 35-44.

GLUCKSMAN J., WEST G. and BERRA T.M. (1976) The introduced fishes of Papua New Guinea with special reference to *Tilapia mossambica*. *Biol. Conserv.* **9**, 37-44.

GREENWOOD P.H. (1974) A review of the family Centropomidae (Pisces: Perciformes). *Bull. Br. Mus. (Nat. Hist.) Zool.* **29**(1), 1-81.

GRIGG G.C. (1965) Studies on the Queensland lungfish, *Neoceratodus forsteri* (Krefft). I. Anatomy, histology, and functioning of the lung. *Aust. J. Zool.* **13**, 243-53.

GRIGG G.C. (1965) Studies on the Queensland lungfish, *Neoceratodus forsteri* (Krefft). II. Thermal acclimation. *Aust. J. Zool.* **13**, 407-11.

GRIGG G.C. (1965) Studies on the Queensland lungfish, *Neoceratodus forsteri* (Krefft). III. Aerial respiration in relation to habits. *Aust. J. Zool.* **13**, 413-21.

HARDISTY M.W. and POTTER I.C. (1971) *The Biology of Lampreys.* Vols I & II. Academic Press, London.

HOBBS D.F. (1948) Trout fisheries in New Zealand. Their development and management. *Fish Bull. N.Z.* **9**, 1–175.

HUGHES R.L. and POTTER I.C. (1969) Studies on gametogenesis and fecundity in the lampreys *Mordacia praecox* and *M. mordax* (Petromyzonidae). *Aust. J. Zool.* **17**, 447–64.

JACKSON P.D. (1975) *Bionomics of brown trout* (Salmo trutta *Linnaeus, 1759*) *in a Victorian Stream with notes on interactions with native fishes.* Ph.D. Thesis, Monash University.

JACKSON P.D. (1976) A note on the food of the Australian grayling, *Prototroctes maraena* Günther (Galaxioidei: Prototroctidae). *Aust. J. Mar. Freshwat. Res.* **27**, 525–8.

JACKSON P.D. (1978) Spawning and early development of the river blackfish, *Gadopsis marmoratus* Richardson (Gadopsiformes: Gadopsidae), in the McKenzie River, Victoria. *Aust. J. Mar. Freshwat. Res.* **29**, 293–8.

JACKSON P.D. (1978) Benthic invertebrate fauna and feeding relationships of brown trout, *Salmo trutta* Linnaeus, and river blackfish, *Gadopsis marmoratus* Richardson, in the Aberfeldy River, Victoria. *Aust. J. Mar. Freshwat. Res.* **29**, 725–42.

JACKSON P.D. (1980) Movement and home range of brown trout, *Salmo trutta* Linnaeus, in the Aberfeldy River, Victoria. *Aust. J. Mar. Freshwat. Res.* **31**, 837–45.

JACKSON P.D. and WILLIAMS W.D. (1980) Effects of brown trout, *Salmo trutta* L., on the distribution of some native fishes in three areas of southern Victoria. *Aust. J. Mar. Freshwat. Res.* **31**, 61–7.

JOHANSEN K., LENFANT C. and GRIGG G.C. (1967) Respiratory control in the lungfish, *Neoceratodus forsteri* (Krefft). *Comp. Biochem. Physiol.* **20**, 835–54.

KOWARSKY J. and ROSS A.H. (1981) Fish movement upstream through a central Queensland (Fitzroy River) coastal fishway. *Aust. J. Mar. Freshwat. Res.* **32**, 93–109.

LAKE J.S. (1967) Rearing experiments with five species of Australian freshwater fishes. I. Inducement to spawning. *Aust. J. Mar. Freshwat. Res.* **18**, 137–53.

LAKE J.S. (1971) *Freshwater Fishes and Rivers of Australia.* Nelson, Sydney.

LAKE J.S. (1978) *Australian Freshwater Fishes.* Nelson, Melbourne.

LLEWELLYN L.C. (1974) Spawning, development, and temperature tolerance of the spangled perch, *Madigania unicolor* (Gunther), from inland waters in Australia. *Aust. J. Mar. Freshwat. Res.* **24**, 73–94.

MACDONALD C.M. (1979) Morphological and biochemical systematics of Australian freshwater and estuarine percichthyid fishes. *Aust. J. Mar. Freshwat. Res.* **29**, 667–98.

MACKAY N.J. (1973) Histological changes in the ovaries of the golden perch, *Plectroplites ambiguus,* associated with the reproductive cycle. *Aust. J. Mar. Freshwat. Res.* **24**, 95–101.

MACLEAN J.L. (1975) *The Potential of Aquaculture in Australia.* Australian Government Publishing Service, Canberra.

MCDOWALL R.M. (1968) The application of the terms anadromous and catadromous to the southern hemisphere salmonoid fishes. *Copeia* **1968**(4), 176–8.

MCDOWALL R.M. (1976) Fishes of the family Prototroctidae (Salmoniformes). *Aust. J. Mar. Freshwat. Res.* **27**, 641–59.

MCDOWALL R.M. (1980) Freshwater fishes and plate tectonics in the southwest Pacific. *Palaeogeog. Palaeoclim. Palaeoecol.* **31**, 337–51.

McDOWALL R.M. (ed.) (1980) *Freshwater Fishes of South-eastern Australia (New South Wales, Victoria and Tasmania).* Reed, Sydney.

McDOWALL R.M. (1981) The relationships of Australian freshwater fishes. In A. Keast (ed.) *Ecological Biogeography of Australia.* Junk, The Hague.

McDOWALL R.M. and FRANKENBERG R.S. (1981) The galaxiid fishes of Australia. *Rec. Aust. Mus.* **33**(10), 443–605.

MEES G.F. (1962) The subterranean freshwater fauna of Yardie Creek Station, North West Cape, Western Australia. *J. Proc. R. Soc. West. Aust.* **45**, 24–32.

MORRISSY N.M. (1972) An investigation into the status of introduced trout. *Report No. 10, Dept. Fisheries and Fauna, Western Australia.*

MORRISSY N.M. (1973) Comparison of strains of *Salmo gairdneri* Richardson from New South Wales, Victoria and Western Australia. *Aust. Soc. Limnol. Bull.* **5**, 11–20.

MORRISSY N.M. (1980) Aquaculture. In W.D. Williams (ed.) *An Ecological Basis for Water Resource Management.* Australian National University Press, Canberra.

MYERS G.S. (1965) *Gambusia*, the fish destroyer. *Aust. Zool.* **13**, 102.

NICHOLLS A.G. (1958) The population of a trout stream and the survival of released fish. *Aust. J. Mar. Freshwat. Res.* **9**, 319–50.

PARRISH R.H. (1966) *A Review of the Gadopsidae, with a description of a new species from Tasmania.* M.Sc. Thesis, Oregon State University.

POLLARD D.A. (1971) The biology of a landlocked form of the normally catadromous salmoniform fish *Galaxius maculatus* (Jenyns). I. Life cycle and origin. *Aust. J. Mar. Freshwat. Res.* **22**, 91–123.

POLLARD D.A. (1974) The freshwater fishes of the Alligator rivers 'Uranium Province' area (top end, Northern Territory), with particular reference to the Magela Creek Catchment (East Alligator River system). Report I. In N.R. Conway, D.R. Davy, M.S. Giles and P.J.F. Newton (eds), and D.A. Pollard *The Alligator Rivers Area Fact Finding Study,* AAEC/E305, 71 pp.

POLLARD D.A. LLEWELLYN L.C. and TILZEY R.D.J. (1980) Management of freshwater fish and fisheries. In W.D. Williams (ed.) *An Ecological Basis for Water Resource Management.* Australian National University Press, Canberra.

POTTER I.C. (1968) *Mordacia praecox*, n.sp., a nonparasitic lamprey (Petromyzonidae), from New South Wales, Australia. *Proc. Linn. Soc. N.S.W.* **92**, 254–61.

POTTER, I.C. (1970) The life cycles and ecology of Australian lampreys of the genus *Mordacia. J. Zool., Lond.* **161**, 487–511.

POTTER I.C., HILLIARD R.W. and BIRD D.J. (1980) Metamorphosis in the southern hemisphere lamprey, *Geotria australis. J. Zool., Lond.* **190**, 405–30.

POTTER I.C., PRINCE P.A. and CROXALL J.P. (1979) Data on the adult marine and migratory phases in the life cycle of the southern hemisphere lamprey, *Geotria australis* Gray. *Environ. Biol. Fishes* **4**(1), 65–9.

POTTER I.C. and STRAHAN R. (1968) The taxonomy of the lampreys *Geotria* and *Mordacia* and their distribution in Australia. *Proc. Linn. Soc. Lond.* **179**, 229–40.

PRIBBLE H.J. (1980) *Annual Report. Carp Program. 1979/1980.* No. **7**. Fisheries and Wildlife Division, Ministry for Conservation, Victoria, Melbourne.

REYNOLDS L.F. and MOORE R. (undated) Comments on the proposed introduction of the Nile Perch (*Lates niloticus* (L.)) to tropical Australian waters. Unpublished report from the Department of Agriculture, Stock and Fisheries, Papua New Guinea.

SENATE STANDING COMMITTEE ON NATIONAL RESOURCES (1979) *The Adequacy of Quarantine.* Australian Government Publishing Service, Canberra.

SHEARER K.D. and MULLEY J.C. (1978) The introduction and distribution of the carp, *Cyprinus carpio* Linnaeus, in Australia. *Aust. J. Mar. Freshwat. Res.* **29**, 551-63.

STOTT B. (1977) On the question of the introduction of the grass carp (*Ctenopharyngodon idella* Val.) into the United Kingdom. *Fish Mgmt* **8**(3), 63-71.

THOMSON J.M. (1974) *Fish of the Ocean and Shore.* Collins, Sydney.

TILZEY R.D.J. (1977) Key factors in the establishment and success of trout in Australia. In D. Anderson (ed.) *Exotic Species in Australia — their Establishment and Success. Proc. Ecol. Soc. Aust.* **10**, 97-105.

TILZEY R.D.J. (1980) Introduced fish. In W.D. Williams (ed.) *An Ecological Basis for Water Resource Management.* Australian National University Press, Canberra.

TRENDALL J.T. and JOHNSON M.S. (1981) Identification by anatomy and gel electrophoresis of *Phalloceros caudimaculatus* (Poeciliidae), previously mistaken for *Gambusia affinis holbrooki* (Poeciliidae). *Aust. J. Mar. Freshwat. Res.* **32**, 993-6.

WEATHERLEY A.H. (1959) Some features of the biology of the tench *Tinca tinca* (Linnaeus) in Tasmania. *J. Anim. Ecol.* **28**, 73-87.

WEATHERLEY A.H. (1963) Zoogeography of *Perca fluviatilis* (Linnaeus) and *Perca flavescens* (Mitchill) with special reference to the effects of high temperature. *Proc. Zool. Soc. Lond.* **141**, 557-76.

WEATHERLEY A.H. (1974) Introduced freshwater fish. In W.D. Williams (ed.) *Biogeography and Ecology in Tasmania.* Junk, The Hague.

WEATHERLEY A.H. (1977) *Perca fluviatilis* in Australia. Zoogeographic expression of a life cycle in relation to an environment. *J. Fish. Res. Bd Can.* **34**, 1464-6.

WEATHERLEY A.H. and COGGER B.M.G. (1977) Fish culture: problems and prospects. *Science, N.Y.* **197**, 427-30.

WEATHERLEY A.H. and LAKE J.S. (1967) Introduced fish species in Australian inland waters. In A.H. Weatherley (ed.) *Australian Inland Waters and their Fauna: Eleven Studies.* Australian National University Press, Canberra.

WHARTON J.C.F. (1973) Spawning induction, artificial fertilization and pond culture of the Macquarie perch (*Macquaria australasica* [Cuvier, 1830]). *Aust. Soc. Limnol. Bull.* **5**, 43-65.

WHITLEY G.P. (1959) The freshwater fishes of Australia. In A. Keast, R.L. Crocker and C.S. Christian (eds) *Biogeography and Ecology in Australia.* Junk, The Hague.

WHITLEY G.P. (1970) Rains of fishes in Australia. *Aust. Nat. Hist.* **17**, 154-9.

WILLIAMS, W.D. (1970) On the proposed introduction of *Lates niloticus* (L.) to Australia. *Aust. Soc. Limnol. Bull.* **2**, 33-5.

WILLIAMS W.D. (1982) The argument against the introduction of the Nile perch to northern Australian fresh waters. *Search* **13**(3-4), 67-70.

6 AMPHIBIANS AND REPTILES

BURBRIDGE A.A., KIRCH J.A.W. and MAIN A.R. (1974) Relationships within the Chelidae (Testudines: Pleurodira) of Australia and New Guinea. *Copeia* **1974**, 392-409.

CANN J. (1978) *Tortoises of Australia.* Angus and Robertson, Sydney.

CHESSMAN B.C. (1978) Ecological studies of freshwater turtles in south-eastern Australia. Ph.D. Thesis, Monash University.

COGGER H.G. (1975) *Reptiles and Amphibians of Australia.* Reed, Sydney.

COGGER H.G. and HEATWOLE H. (1981) The Australian reptiles: origins, biogeography, distribution patterns and island evolution. In A. Keast (ed.) *Ecological Biogeography of Australia*. Junk, The Hague.

GLASS M. and JOHANSEN K. (1976) Control of breathing in *Acrochordus javanicus*, an aquatic snake. *Physiol. Zool.* **49**, 328–40.

GOODE J. (1967) *Freshwater Tortoises of Australia and New Guinea (in the Family Chelidae)*. Lansdowne Press, Melbourne.

GOODE J. and RUSSELL J. (1968) Incubation of eggs of three species of chelid tortoises, and notes on their embryological development. *Aust. J. Zool.* **16**, 749–61.

GRIGG G.C. and ALCHIN J. (1976) The role of the cardiovascular system in thermo-regulation of *Crocodylus johnstoni*. *Physiol Zool.* **49**, 24–36.

GUGGISBERG C.A.W. (1972) *Crocodiles. Their Natural History, Folklore and Conservation*. David and Charles, Newton Abbot.

HEATWOLE H. and SEYMOUR R.S. (1975) Pulmonary and cutaneous oxygen uptake in sea snakes and a file snake. *Comp. Biochem. Physiol.* **51A**, 399–405.

HEYER W.R. and LIEM D.S. (1976) Analysis of the intrageneric relationships of the Australian frog family Myobatrachidae. *Smithson. Contrib. Zool.* **233**, 1–29.

JENKINS R.W.G. (1979) The status of endangered Australian reptiles. In M.J. Tyler (ed.) *The Status of Endangered Australasian Wildlife*. Royal Zoological Society of South Australia, Adelaide.

JOHNSON C.R. (1973) Behaviour of the Australian crocodiles, *Crocodylus johnstoni* and *C. porosus*. *Zool. J. Linn. Soc.* **52**, 315–36.

LEE A.K. (1968) Water economy of the burrowing frog, *Heleioporus eyrei* (Gray). *Copeia* **1968**, 741–5.

LEE A.K. and MERCER E.H. (1967) Cocoon surrounding desert-adapted frogs. *Science, N.Y.* **159**, 87–8.

LIEM D.S. (1973) A new genus of frog of the family Leptodactylidae from S.E. Queensland, Australia. *Mem. Qld. Mus.* **16**(3), 459–70.

LITTLEJOHN M.J. (1965) Vocal communication in frogs. *Aust. Nat. Hist.* **15**(2), 52–5.

LITTLEJOHN, M.J. (1968) Frog calls and the species problem. *Aust. Zool.* **14**(3), 259–64.

LITTLEJOHN M.J. (1981) The Amphibia of mesic southern Australia: a zoogeographic perspective. In A. Keast (ed.) *Ecological Biogeography of Australia*. Junk, The Hague.

MAGNUSSON W.E. (1980) Habitat required for nesting by *Crocodylus porosus* (Reptilia: Crocodilidae) in Northern Australia. *Aust. J. Wildl. Res.* **7**, 149–56.

MAIN A.R. (1968) Ecology, systematics and evolution of Australian frogs. *Adv. Ecol. Res.* **5**, 37–86.

MARTIN A.A. (1967) Australian anuran life histories: some evolutionary and ecological aspects. In A.H. Weatherley (ed.) *Australian Inland Waters and their Fauna: Eleven Studies*. Australian National University Press, Canberra.

MARTIN A.A. and TYLER M.J. (1978) The introduction into Western Australia of the frog *Limnodynastes tasmaniensis* Gunther. *Aust. Zool.* **19**(3), 321–5.

McCOY F. (1884) *Physiognathus lesueri* (Gray), var. *Howitti* (McCoy). The Gippsland water lizard. *Prodromus, Zoology of Victoria* **9**, 7–10.

MESSEL H. (1980) Rape of the north. *Habitat* **8**(2), 3–6.

MESSEL H., WEBB, G.J.W., YERBURY M. and GRIGG G.C. (1977) A study of *Crocodylus porosus* in Northern Australia. In H. Messel and S.T. Butler (eds) *Australian Animals and their Environment*. Shakespeare Press, Sydney.

MOORE J.A. (1961) The frogs of eastern New South Wales. *Bull. Amer. Mus. Nat. Hist.* **121**(3), 151–385.

PENGILLEY R.K. (1971) The food of some Australian anurans (Amphibia). *J. Zool. Lond.* **163**, 93–103.

RAWLINSON P.A. (1974) Biogeography and ecology of the reptiles of Tasmania and the Bass Strait area. In W.D. Williams (ed.) *Biogeography and Ecology in Tasmania.* Junk, The Hague.

SCHODDE R., MASON I. and WOLFE T.O. (1972) Further records of the pitted-shelled turtle (*Carettochelys insculpta*) from Australia. *Trans. Roy. Soc. S.A.* **96**, 115–7.

TAYLOR J.A. (1979) The foods and feeding habits of subadult *Crocodylus porosus* Schneider in northern Territory. *Aust. Wildl. Res.* **6**, 347–59.

TYLER M.J. (1978) *Amphibians of South Australia.* Government Printer, Adelaide.

TYLER M.J. (1979) The impact of European man upon Australasian amphibians. In M.J. Tyler (ed.) *The Status of Endangered Australasian Wildlife.* Royal Zoological Society of South Australia, Adelaide.

TYLER M.J. (1980) Introduced amphibians: the cane toad. In W.D. Williams (ed.) *An Ecological Basis for Water Resource Management.* Australian National University Press, Canberra.

TYLER M.J. (1982) *Frogs.* 2nd edn. Collins, Sydney.

TYLER M.J. (ed.) (1983) *The Gastric Brooding Frog.* Croom Helm, Beckenham.

TYLER M.J. and CROOK G.A. (1980) *Frogs of the Magela Creek System, Alligator Rivers Region, Northern Territory, Australia.* Report to the Office of the Supervising Scientist, Alligator Rivers Region, Adelaide.

TYLER M.J., DAVIES M. and MARTIN A.A. (1981) Australian frogs of the leptodactylid genus *Uperoleia* Gray. *Aust. J. Zool., Suppl. Ser.* **79**, 1–64.

TYLER M.J., ROBERTS J.D. and DAVIES M. (1980) Field observations on *Arenophryne rotunda* Tyler, a leptodactylid frog of coastal sandhills. *Aust. Wildl. Res.* **7**, 295–304.

TYLER M.J., WATSON G.F. and MARTIN A.A. (1981) The Amphibia: diversity and distribution. In A. Keast (ed.) *Ecological Biogeography of Australia.* Junk, The Hague.

WEBB G.J.W. and MESSEL H. (1979) Wariness in *Crocodylus porosus. Aust. Wildl. Res.* **6**, 227–34.

WEBB G.J.W., MESSEL H. and MAGNUSSON W.E. (1977) The nesting of *Crocodylus porosus* in Arnhem Land, Northern Australia. *Copeia,* **1977**(2), 238–49.

7 WATERBIRDS

BRAITHWAITE L.W. (1975) Managing waterfowl in Australia. *Proc. Ecol. Soc. Aust.* **8**, 107–28.

BRAITHWAITE L.W. (1976) Notes on the breeding of the freckled duck in the Lachlan River Valley. *Emu* **76**, 127–32.

BRAITHWAITE L.W. (1976) Environment and timing of reproduction and flightlessness in two species of Australian ducks. *Proc. 16th Intern. Ornith. Congr.* 489–501.

BRAITHWAITE L.W. (1976) Breeding seasons of waterfowl in Australia. *Proc. 16th Intern. Ornith. Congr.* 235–47.

BRAITHWAITE L.W. (1977) Ecological studies of the black swan. 1. The egg, clutch and incubation. *Aust. Wildl. Res.* **4**, 59–79.

BRAITHWAITE L.W. (1980) Waterfowl resources and their management. In W.D. Williams (ed.) *An Ecological Basis for Water Resource Management.* Australian National University Press, Canberra.

BRAITHWAITE L.W. and FRITH H.J. (1969) Waterfowl in an inland swamp in New South Wales. I. Habitat. *CSIRO Wildl. Res.* **14**, 1–16.

BRAITHWAITE L.W. and FRITH H.J. (1969) Waterfowl in an inland swamp in New South Wales. II. Breeding. *CSIRO Wildl. Res.* **14**, 65–109.

BRAITHWAITE L.W. and MILLER B. (1975) The mallard, *Anas platyrhynchos*, and mallard-black duck, *Anas superciliosa rogersi*, hybridization. *Aust. Wildl. Res.* **2**, 47–61.

BRAITHWAITE L.W. and STEWART D.A. (1975) Dynamics of water bird populations on the Alice Springs sewage farm, N.T. *Aust. Wildl. Res.* **2**, 85–90.

BRIGGS, S.V. (1977) Variation in waterbird numbers at four swamps on the northern Tablelands of New South Wales. *Aust. Wildl. Res.* **4**, 301–9.

CARRICK R. (1959) The food and feeding habits of the straw-necked ibis *Threskiornis spinicollis* (Jameson), and the white ibis *T. molucca* (Cuvier), in Australia. *CSIRO Wildl. Res.* **4**, 69–92.

CARRICK R. (1962) Breeding, movements and conservation of ibises (Threskiornithidae) in Australia. *CSIRO Wildl. Res.* **7**, 71–88.

COWAN I.M. (1973) The conservation of Australian waterfowl. *A.F.A.C. Special Publication No. 2,* Australian Government Publishing Service, Canberra.

COWLING S.J. (1977) Classification of the wetland habitats of waterbirds. *Aust. Mar. Sci. Bull.* **58**, 15–16.

COWLING S.J. (1979) The status of endangered waterfowl and wetlands in Australia. In M.J. Tyler (ed.) *The Status of Endangered Australasian Wildlife.* Royal Zoological Society of South Australia, Adelaide.

CRAWFORD D.N. (1979) Waterbirds: indices and fluctuations in dry-season refuge areas, Northern Territory. *Aust. Wildl. Res.* **6**, 97–103.

DORWARD D.F. (1977) *Wild Australia. A View of Birds and Men.* Collins, Sydney.

FRITH H.J. (1959) The ecology of wild ducks in inland New South Wales. I. Waterfowl habitats. II. Movements. III. Food habits. IV. Breeding. *CSIRO Wildl. Res.* **4**, 97–181.

FRITH H.J. (1965) Ecology of the freckled duck, *Stictonetta naevosa* (Gould). *CSIRO Wildl. Res.* **10**, 125–39.

FRITH H.J. (1977) *Waterfowl in Australia.* 2nd edn. Reed, Sydney.

FRITH H.J., BRAITHWAITE L.W. and McKEAN J.L. (1969) Waterfowl in an inland swamp in New South Wales. II. Food. *CSIRO Wildl. Res.* **14**, 17–64.

FRITH, H.J. and DAVIES S.J.J.F. (1961) Ecology of the magpie goose, *Anseranas semipalmata* Latham (Anatidae). *CSIRO Wildl. Res.* **6**, 91–141.

GARNETT S.T. (1978) The behaviour patterns of the dusky moorhen, *Gallinula tenebrosa* Gould (Aves: Rallidae). *Aust. Wildl. Res.* **5**, 363–84.

GARNETT S.T. (1980) The social organization of the dusky moorhen, *Gallinula tenebrosa* Gould (Aves: Rallidae). *Aust. Wildl. Res.* **7**, 103–12.

GOODRICK G.N. (1979) Food of the black duck and grey teal in coastal northern New South Wales. *Aust. Wildl. Res.* **6**, 319–24.

HALE W.G. (1980) *Waders.* Collins, London.

KEAST A. (1981) The evolutionary biogeography of Australian birds. In A. Keast (ed.) *Ecological Biogeography of Australia.* Junk, The Hague.

LACK D. (1974) *Evolution Illustrated by Waterfowl.* Blackwell Scientific Publications, Oxford.

LAVERY H.J. (1970) The comparative ecology of waterfowl in north Queensland. *Wildfowl* 21, 69–77.

LAVERY H.J. and BLACKMAN J.G. (1969) The cranes of Australia. *Qld. Agric. J.* 95, 156–62.

McKILLIGAN N.G. (1975) Breeding and movements of the straw-necked ibis in Australia. *Emu* 75, 199–212.

MILLER A.H. (1963) The fossil flamingos of Australia. *Condor* 65, 289–99.

MILLER B. (1979) Ecology of the little black cormorant, *Phalacrocorax sulcirostris*, and little pied cormorant, *P. melanoleucos*, in inland New South Wales. I. Food and feeding habits. *Aust. Wildl. Res.* 6, 79–95.

MISSEN R. and TIMMS B.V. (1974) Seasonal fluctuations in waterbird populations on three lakes near Camperdown, Victoria. *Aust. Birdwatcher* 5, 128–35.

MURRAY M.D. and CARRICK R. (1964) Seasonal movements and habitats of the silver gull, *Larus novaehollandiae* Stephens, in south-eastern Australia. *CSIRO Wildl. Res.* 9, 160–88.

OLSEN P., SETTLE H. and SWIFT R. (1980) Organochlorine residues in wings of ducks in south-eastern Australia. *Aust. Wildl. Res.* 7, 139–47.

REID A.J., SHAW N.J. and WHEELER W.R. (1972) *Birds of Victoria. Inland Waters.* Gould League of Victoria, Melbourne.

RIDPATH M.G. (1972) The Tasmanian native hen, *Tribonyx mortierii*. I. Patterns of behaviour. II. The individual, the group, and the population. III. Ecology. *CSIRO Wildl. Res.* 17, 1–118.

ROWLEY I. (1975) *Bird Life.* Collins, Sydney.

SLATER P. (1970) *A Field Guide to Australian Birds: Non-Passerines.* Rigby, Adelaide.

VESTJENS W.J.M. (1975) Feeding behaviour of spoonbills at Lake Cowal, N.S.W. *Emu* 75, 132–6.

VESTJENS W.J.M. (1977) Breeding behaviour and ecology of the Australian pelican, *Pelicanus conspicillatus*, in New South Wales. *Aust. Wildl. Res.* 4, 37–58.

WOOLLER R.D. and DUNLOP J.N. (1979) Multiple laying by the silver gull, *Larus novaehollandiae* Stephens, on Carnac Island, Western Australia. *Aust. Wildl. Res.* 6, 325–35.

8 MAMMALS

ANGEL L.M. (1967) The life-cycle of *Echinoparyphium hydromyos* sp.nov. (Digenea: Echinostomatidae) from the Australian water-rat. *Parasitology* 57, 19–30.

BARRETT C. (1944) *The Platypus.* Robertson and Mullens, Melbourne.

BARROW G.J. (1964) *Hydromys chrysogaster*—some observations. *Qld Nat.* 17, 43–4.

BURRELL H. (1974) *The Platypus.* Rigby, Adelaide. First published 1927.

BRYAN R.P., BAINBRIDGE M.J. and KERR J.D. (1976) A study of helminth parasites in the gastrointestinal tract of the swamp buffalo, *Bubalus bubalis* Lydekker, in the Northern Territory. *Aust. J. Zool.* 24, 417–21.

CALABY J.H. (1968) The platypus (*Ornithorhynchus anatinus*) and its venomous characteristics. In W. Bücherl, E. Buckley and V. Deulofeu (eds) *Venomous Animals and Their Venoms.* Academic Press, New York.

CALDWELL W.H. (1887) The embryology of Monotremata and Marsupialia, Part I. *Phil. Trans. Roy. Soc. Lond.*, Series B. 179, 463–86.

CARRICK F.N. and HUGHES R.L. (1978) Reproduction in male monotremes. *Aust. Zool.* 20(1), 211–31.

COLLINS D. (1802) *An Account of the English Colony in New South Wales.* **2**, 321-8.

COSTELLO P. (1974) *In Search of Lake Monsters.* Garnstone Press, London.

FAHIMUDDIN M. (1975) *Domestic Water Buffalo.* Oxford and IBH Publishing Co., New Delhi.

FANNING F.D. and DAWSON T.J. (1980) Body temperature variability in the Australian water rat, *Hydromys chrysogaster*, in air and water. *Aust. J. Zool.* **28**, 229-38.

FARAGHER R.A., GRANT T.R. and CARRICK F.N. (1979) The food of the platypus *Ornithorhynchus anatinus* (Shaw) in the upper Shoalhaven River, New South Wales. *Aust. J. Ecol.* **4**, 171-9.

FLEAY D. (1964) The rat that mastered the waterways. *Wildlife in Australia* **1**, 3-7.

GRANT T.R. and CARRICK F.N. (1978) Some aspects of the ecology of the platypus, *Ornithorhynchus anatinus*, in the upper Shoalhaven River, New South Wales. *Aust. Zool.* **20**(1), 181-99.

GRANT T.R. and DAWSON T.J. (1978) Temperature regulation in the platypus (*Ornithorhynchus anatinus*): maintenance of body temperature in air and water. *Physiol. Zool.* **51**, 1-6.

GRANT T.R. and DAWSON T.J. (1978) Temperature regulation in the platypus (*Ornithorhynchus anatinus*): production and loss of metabolic heat in air and water. *Physiol. Zool.* **51**, 315-32.

GRANT T.R., WILLIAMS R. and CARRICK F.N. (1977) Maintenance of the platypus (*Ornithorhynchus anatinus*) in captivity under laboratory conditions. *Aust. Zool.* **19**(2), 117-24.

GRIFFITHS M. (1978) *The Biology of the Monotremes.* Academic Press, New York.

GUNN R.C. (1847) On the 'Bunyip' of Australia felix. *Tasmanian J. Sci.* **3**, 147-9.

HEUVELMANS B. (1958) *On the Track of Unknown Animals.* Rupert, Hart-Davis, London.

HOME E.A. (1802) A description of the anatomy of the *Ornithorhynchus paradoxus. Phil. Trans. Roy. Soc. Lond.* **92**, 67-84.

HUGHES R.L. and CARRICK F.N. (1978) Reproduction in female monotremes. *Aust. Zool.* **20**(1), 233-51.

HYETT J. and SHAW N. (1980) *Australian Mammals. A Field Guide for New South Wales, South Australia, Victoria and Tasmania.* Nelson, Melbourne.

JOHANSEN K., LENFANT C. and GRIGG G.C. (1966) Respiratory properties of blood and responses to diving of the platypus, *Ornithorhynchus anatinus* (Shaw) *Comp. Biochem. Physiol.* **18**, 597-608.

KEAST A. (1972) Australian mammals: zoogeography and evolution. In A. Keast, F.C. Erk and B. Glass (eds) *Evolution, Mammals, and Southern Continents.* State University of New York Press, Albany.

LEE A.K., BAVERSTOCK P.R. and WATTS C.H.S. (1981) Rodents—the late invaders. In A. Keast (ed.) *Ecological Biogeography of Australia.* Junk, The Hague.

LETTS G.A. (1964) Feral animals in the Northern Territory. *Aust. Vet. J.* **40**, 84-8.

MAGNUSSON W.E., WEBB G.J.W. and TAYLOR J.A. (1976) Two new locality records, a new habitat and a nest description for *Xeromys myoides* Thomas (Rodentia: Muridae). *Aust. Wildl. Res.* **3**, 153-7.

MARSHALL A.J. (1966) On the disadvantages of wearing fur. In A.J. Marshall (ed.) *The Great Extermination.* Heinemann, London.

MCNALLY J. (1960) The biology of the water rat *Hydromys chrysogaster* Geoffroy (Muridae: Hydromyinae) in Victoria. *Aust. J. Zool.* **8**, 170-80.

NEW SOUTH WALES FAUNA PROTECTION PANEL (1954) The Platypus in New South

Wales: Report on a Survey of the New South Wales distribution of the Platypus (*Ornithorhynchus anatinus*). Government Printer, Sydney.

REDHEAD T.D. and McKEAN J.L. (1974) A new record of the false water-rat, *Xeromys myoides*, from the Northern Territory of Australia. *Aust. Mammal.* **1**, 347–54.

RIDE, W.D.L. (1970) *A Guide to the Native Mammals of Australia*. Oxford University Press, Melbourne.

SHAW, G. (1799) The duck-billed platypus. *The Naturalists' Miscellany* **10**.

SIMPSON G.G. (1961) Historical zoogeography of Australian mammals. *Evolution* **15**, 431–46.

SMYTH D.M. (1973) Temperature regulation in the platypus *Ornithorhynchus anatinus* (Shaw) *Comp. Biochem. Physiol.* **45A**, 705–15.

STOCKER G.C. (1970) The effects of water buffaloes on paperbark forests in the Northern Territory. *Aust. For. Res.* **5**(1), 39–34.

STRAHAN R. and THOMAS D.E. (1975) Courtship of the platypus, *Ornithorhynchus anatinus*. *Aust. Zool.* **18**(3), 165–78.

SOUTH D. (1971) *Ornithorhynchus* or the platypus. A bibliography of works in the English language. Unpublished manuscript. Australian Museum Library, Sydney.

TEMPLE-SMITH P.D. (1973) *Seasonal Breeding Biology of the Platypus*, Ornithorhynchus anatinus *(Shaw, 1799), with Special Reference to the Male*. Ph.D. Thesis, Australian National University, Canberra.

TROUGHTON E. LE G. (1941) Australian water-rats: their origin and habits. *Aust. Mus. Mag.* **7**, 377–81.

TULLOCH D.G. (1967) *The Distribution, Density and Social Behaviour of the Water Buffalo in the Northern Territory*. M.Sc. (Agric.) Thesis, University of Queensland, Brisbane.

TULLOCH D.G. (1969) Home range in feral buffalo. *Aust. J. Zool.* **17**, 143–52.

TULLOCH D.G. (1970) Seasonal movements and distribution of the sexes in the water buffalo, *Bubalus bubalis*, in the Northern Territory. *Aust. J. Zool.* **18**, 399–414.

TULLOCH D.G. (1975) Buffalo in the northern swamp lands. In R.L. Reid (ed.) *Proceedings of the III World Conference on Animal Production, Melbourne*. Sydney University Press, Sydney.

TULLOCH D.G. (1978) The water buffalo, *Bubalus bubalis*, in Australia: grouping and home range. *Aust. Wildl. Res.* **5**, 327–54.

WATTS C.H.S. and ASLIN H.J. (1981) *The Rodents of Australia*. Angus and Robertson, Sydney.

WHITLEY G. (1940) Mystery animals of Australia. *Aust. Mus. Mag.* March, 132–9.

WILLIAMS F.J. (1884) Notes on the platypus. *Vict. Nat.* **1**, 87–9.

WOOLLARD P., VESTJENS W.J.M. and MACLEAN L. (1978) The ecology of the eastern water rat *Hydromys chrysogaster* at Griffith, N.S.W.: food and feeding habits. *Aust. Wildl. Res.* **5**, 59–73.

9 MICROSCOPIC PLANTS

BELCHER H. and SWALE E. (1976) *A Beginner's Guide to Freshwater Algae*. HMSO, London.

BREMER H.J. (1978) Hazards and problems in the utilization of microalgae for human nutrition. *Arch. Hydrobiol. Beih. Ergebn. Limnol.* **11**, 218–22.

BONEY A.D. (1975) *Phytoplankton.* Studies in Biology no. 52. Edward Arnold, London.

BOROWITZKA L.J. (1981) The microflora. Adaptations to life in extremely saline lakes. In W.D. Williams (ed.) *Salt Lakes.* Junk, The Hague.

BROCK T.D. (1979) *Biology of Microorganisms.* 3rd edn. Prentice-Hall, London.

BURNS F.L. and POWLING I.J. (eds) (1981). *Destratification of Lakes and Reservoirs to Improve Water Quality.* Australian Government Publishing Service, Canberra.

CARMICHAEL J.W., KENDRICK W.B., CONNERS I.L. and SIGLER L. (1980) *Genera of Hyphomycetes.* University of Alberta Press, Edmonton.

COLLINS M. (1978) Algal toxins. *Miocrobiol. Rev.* **42**, 725–46.

COWLING S.W. and WAID J.S. (1963) Aquatic Hyphomycetes in Australia. *Aust. J. Sci.* **26**(4), 122–3.

CROOME R.L. and TYLER P.A. (1975) Phytoplankton biomass and primary productivity of Lake Leake and Tooms Lake, Tasmania. *Hydrobiologia* **46**, 435–43.

DICK M.W. (1976) The ecology of aquatic Phycomycetes. In E.B.G. Jones (ed.) *Recent Advances in Aquatic Mycology.* Elek Science, London.

DOETSCH R.N. and COOK T.M. (1973) *Introduction to Bacteria and their Ecobiology.* Medical and Technical Publishing, Lancaster.

FJERDINGSTAD E. (1979) Sulphur bacteria. *ASTM Spec. Tech. Publ.* **650**, 1–121.

FOGG G.E. (1975) *Algal Cultures and Phytoplankton Ecology,* 2nd edn. University of Wisconsin Press, Madison and Milwaukee.

FOGG G.E., STEWART W.D.P., FAY P. and WALSBY A.E. (1973) *The Blue-green Algae.* Academic Press, London.

GANF G.G. (1980) Factors controlling the growth of phytoplankton in Mount Bold Reservoir, South Australia. *AWRC Tech. Paper* No. 48.

GANF G.G. (1980) Ecological considerations in the management of reservoir phytoplankton. In W.D. Williams (ed.) *An Ecological Basis for Water Resource Management.* Australian National University Press, Canberra.

GROBBELAAR J.U., SOEDER C.J. and TOERIEN D.F. (eds) (1981) *Wastewater for Aquaculture.* University of the O.F.S. Publication, Series C, No. 3, Bloemfontein.

GUTTERIDGE, HASKINS & DAVEY PTY LTD (1977) *Planning for the Use of Sewage. Summary Report.* Australian Government Publishing Service, Canberra.

HAPPEY-WOOD C.M. (1976) The occurrence and relative importance of nannochlorophyta in freshwater algal communities. *J. Ecol.* **64**, 279–92.

INGOLD C.T. (1975) An illustrated guide to aquatic and water-borne Hyphomycetes (Fungi Imperfecti) with notes on their biology. *Freshw. Biol. Assoc. Sci. Publ.* No. **30**, 1–96.

JACKSON D.F. (ed.) (1962) *Algae and Man.* Plenum Press, New York.

JEFFREY J.M. and WILLOUGHBY L.G. (1964) A note on the distribution of *Allomyces* in Australia. *Nova Hedwigia* **7**, 505–15.

LING H.U. and TYLER P.A. (1980) *Freshwater Algae of the Alligator Rivers Region, Northern Territory of Australia.* A Report to the Office of the Supervising Scientist; Botany Department, University of Tasmania, Hobart.

MAY V. (1972) Blue-green algal blooms at Braidwood New South Wales. *Sci. Bull. No.* **82**, New South Wales Dept. Agric., Sydney.

MITCHELL B.D. and WILLIAMS W.D. (1982) *The Performance of Tertiary Treatment Ponds and the Role of Algae, Macrophytes, and Zooplankton in the Waste Treatment Process.* Australian Government Publishing Service, Canberra.

MORRIS I. (ed.) (1980) *The Physiological Ecology of Phytoplankton.* Blackwell Scientific Publications, Oxford.

PATERSON R.A. (1971) Lacustrine fungal communities. In J. Cairns Jr (ed.) *The Structure and Function of Fresh-water Microbial Communities. Res. Div. Monogr.* **3**. Virginia Polytechnic Institute and State University, Blacksburg, Virginia.

PATRICK R. (1973) Use of algae, especially diatoms, in the assessment of water quality. In J. Cairns Jr and K.L. Dickson (eds) *Biological Methods for the Assessment of Water Quality.* American Society for Testing and Materials, Philadelphia.

PLAYFAIR G.I. (1915) Freshwater algae of the Lismore district with an appendix on the algal fungi and Schizomycetes. *Proc. Linn. Soc. N.S.W.* **40**(2), 309-62.

PFENNIG N. (1978) General physiology and ecology of photosynthetic bacteria. In R.K. Clayton and W. R. Sistrom (eds) *The Photosynthetic Bacteria.* Plenum, New York.

POTTER I.C., CANNON D. and MOORE J.W. (1975) The ecology of algae in the Moruya River, Australia. *Hydrobiologia* **47**, 415-30.

REYNOLDS C.S. and WALSBY A.E. (1975) Water blooms. *Biol. Rev.* **50**, 437-81.

RHEINHEIMER G. (1971) *Aquatic Microbiology.* Wiley, London.

ROHDE W., LIKENS G.E. and SERRUYA C. (eds) (1979) *Lake Metabolism and Management.* Schweizerbart'sche Verlagsbuchhandlung, Stuttgart.

ROUND F.E. (1973) *The Biology of the Algae.* 2nd edn. Edward Arnold, London.

SANDBANK E. and HEPHER B. (1978) The utilization of microalgae as a feed for fish. *Arch. Hydrobiol. Beih. Ergebn. Limnol.* **11**, 108-20.

SHAW D.E. (1972) *Ingoldiella hamata* gen. et sp. nov., a fungus with clamp connexions from a stream in north Queensland. *Trans. Br. Mycol. Soc.* **59**, 255-59.

SHELEF G., ORON G. and MORAINE R. (1978) Economic aspects of microalgae production on sewage. *Arch. Hydrobiol. Beih. Ergebn. Limnol.* **11**, 281-94.

SMALLS I.C. (1980) Algal problems in water supplies. In W.D. Williams (ed.) *An Ecological Basis for Water Resource Management.* Australian National University Press, Canberra.

SKINNER F.A. and SHEWAN J.M. (eds) (1977) *Aquatic Microbiology.* Academic Press, London.

SPARROW F.K. (1960) *Aquatic Phycomycetes.* 2nd edn. University of Michigan Press, Ann Arbor.

STALEY J.T., MARSHALL K.C. and SKERMAN V.B.D. (1980) Budding and prosthecate bacteria from freshwater habitats of various trophic states. *Microbiol. Ecology* **5**, 245-51.

TYLER P.A. (1970) Taxonomy of Australian freshwater algae. I. The genus *Micrasterias* in south-eastern Australia. *Br. phycol. J.* **5**(2), 211-34.

TYLER P.A. (1970) Hyphomicrobia and the oxidation of manganese in aquatic ecosystems. *Antonie van Leeuwenhoek, J. Microbiol. Serol.* **36**, 567-78.

TYLER P.A. and MARSHALL K.C. (1967) Form and function in manganese-oxidising bacteria. *Arch. Mikrobiol.* **56**, 344-53.

VALLENTYNE J.R. (1974) *The Algal Bowl. Lakes and Man.* Department of the Environment, Fisheries and Marine Service, Ottawa.

WAKE L.V. and HILLEN L.W. (1981) Nature and hydrocarbon content of blooms of the alga *Botryococcus braunii* occurring in Australian freshwater lakes. *Aust. J. Mar. Freshwat.* **32**, 353-67.

WALKER K.F. and HILLMAN T.J. (1977) *Limnological Survey of the River Murray in Relation to Albury-Wodonga, 1973-1976.* Albury-Wodonga Development Corporation and Gutteridge, Haskins & Davey, Melbourne.

WALKER K.F. and HILLMAN T.J. (1982) Phosphorus and nitrogen loads in waters associated with the River Murray near Albury-Wodonga, and their effects on phytoplankton populations. *Aust. J. Mar. Freshwat. Res.* **33**, 223-43.

WERNER D. (ed.) (1977) *The Biology of Diatoms.* Blackwell Scientific Publications, Oxford.

WOOD G. (1975) An assessment of eutrophication in Australian inland waters. *A WRC Tech. Paper* No. 15.

10   MACROPHYTES

ANDRES L.A. (1977) The economics of biological control of weeds. *Aquatic Botany* **3**, 111–23.

ANDRES L.A. and BENNETT F.D. (1975) Biological control of aquatic weeds. *Ann. Rev. Entomol.* **20**, 31–46.

ARBER A. (1920) *Water Plants: A Study of Aquatic Angiosperms.* Cambridge University Press, Cambridge.

ASTON H.I. (1973) *Aquatic Plants of Australia.* Melbourne University Press, Melbourne.

ASTON H.I. (1977) Supplement to *Aquatic Plants of Australia.* Melbourne University Press, Melbourne.

BILL S.M. and GRAHAM W.A.E. (1970) *Chemical Weed Control in Irrigation Channels and Drains.* State Rivers and Water Supply Commission, Melbourne.

BOWMER K.H., O'LOUGHLIN E.M., SHAW K. and SAINTY G.R. (1976) Residues of dichlobenil in irrigation water. *J. Environ. Qual.* **5**, 315–19.

BRIGGS S.V. (1981) Freshwater wetlands. In R.H. Groves (ed.) *Australian Vegetation.* Cambridge University Press, Cambridge.

BROCK M.A. (1979) *The Ecology of Salt Lake Hydrophytes. The Synecology of Saline Ecosystems and the Autecology of the Genus* Ruppia *L. in the South-East of South Australia.* Ph.D. Thesis, University of Adelaide.

BROCK M.A. (1981) The ecology of halophytes in the south-east of South Australia. In W.D. Williams (ed.) *Salt Lakes.* Junk, The Hague.

BURNE R.V., BAULD J. and DE DECKKER P. (1980) Saline lake charophytes and their geological significance. *J. Sediment. Petrol.* **50**, 281–93.

CSIRO (1981) Submerged aquatic plants (Topic essay). *CSIRO Division of Irrigation Research Report*, 1978-1980, 18–28.

COOK C.D.K., GUT J.B., RIX E.M., SCHNELLER J. and SEITZ M. (1974) *Water Plants of the World.* Junk, The Hague.

EARDLEY C.M. (1943) An ecological study of the Eight Mile Creek Swamp, a natural South Australian fen formation. *Trans. Roy. Soc. S.A.* **67**, 200–23.

FINLAYSON C.M., FARRELL T.P. and GRIFFITHS D.J. (1982) Treatment of sewage effluent using the water fern *Salvinia. Water Research Foundation of Australia,* Report No. **57**, 1–37.

GANF G.G. (1977) Ecological aspects of water weeds. In *The Menace of Water Hyacinth and Other Aquatic Weeds.* Water Research Foundation of Australia, Adelaide.

GOOD R.E., WHIGHAM D.F. and SIMPSON R.L. (eds) (1978) *Freshwater Wetlands. Ecological Processes and Management Potential.* Academic Press, New York.

HARLEY K.L.S. (1977) Biological control of aquatic weeds. In *The Menace of Water Hyacinth and Other Aquatic Weeds.* Water Research Foundation of Australia, Adelaide.

HASLAM S.M. (1978) *River Plants. The Macrophytic Vegetation of Watercourses.* Cambridge University Press, Cambridge.

HUTCHINSON G.E. (1975) *A Treatise on Limnology. Volume III. Limnological Botany.* Wiley, New York.

KADLEC R.H. and TILTON D.L. (1979) The use of freshwater wetlands as a tertiary wastewater treatment alternative. *Crit. Rev. Environ. Control* **9**(2), 185–212.

KATS N.Y. (1969) [Swamps of Australia and Tasmania.] *Bjull. Mosk. Obsc. Ispyt Prir (Otd. Biol.)* **74**, 106–16. [In Russian].

LITTLE E.C.S. (1981) *Handbook of Utilization of Aquatic Plants.* 2nd edn. FAO, Rome.

MILLINGTON R.J. (1954) *Sphagnum* bogs of the New England plateau, New South Wales. *J. Ecol.* **42**, 328–44.

MITCHELL D.S. (ed.) (1974) *Aquatic Vegetation and Its Use and Control.* UNESCO, Paris.

MITCHELL D.S. (1977) *Aquatic Weeds in Australian Inland Waters.* Australian Government Publishing Service, Canberra.

MITCHELL D.S. (1977) Water weed problems in irrigation systems. In E.B. Wirthington (ed.) *Arid Land Irrigation in Developing Countries.* Pergamon Press, Oxford.

MITCHELL D.S. (1980) Aquatic Weeds. In W.D. Williams (ed.) *An Ecological Basis for Water Resource Management.* Australian National University Press, Canberra.

O'LOUGHLIN E.M. and BOWMER K.H. (1975) Dilution and decay of aquatic herbicides in flowing channels. *J. Hydrol.* **26**, 217–35.

PIETERSE A.H. (1977) Biological control of aquatic weeds: perspective for the tropics. *Aquatic Botany* **3**, 133–41.

ROOM P.M., HARLEY K.L.S., FORNO I.W. and SANDS D.P.A. (1981) Successful biological control of the floating weed salvinia. *Nature, Lond.* **294**, 78–80.

SAINTY G.R. (1973) *Aquatic Plants Identification Guide.* Water Conservation and Irrigation Commission, Sydney.

SAINTY G.R. and JACOBS S.W.L. (1981) *Water Plants of New South Wales.* Water Resources Commission, New South Wales, Sydney.

SCULTHORPE C.D. (1967) *The Biology of Aquatic Vascular Plants.* Edward Arnold, London.

SHIEL R.J. (1976) Associations of entomostraca with weedbed habitats in a billabong of the Goulburn River, Victoria. *Aust. J. Mar. Freshwat. Res.* **27**, 533–49.

SPECHT R.L. (1981) Major vegetation formations in Australia. In A. Keast (ed.) *Ecological Biogeography in Australia.* Junk, The Hague.

WESTLAKE D.F. (1975) Primary production of freshwater macrophytes. In J.P. Cooper (ed.) *Photosynthesis and Productivity in Different Environments.* Cambridge University Press, Cambridge.

WILSON K.L. (1981) A synopsis of the genus *Scirpus* sens. lat. (Cyperaceae) in Australia. *Telopea* **2**(2), 153–72.

WOOD R.D. (1971) Characeae of Australia. *Nova Hedwigia* **22**, 1–120.

11 CONSERVATION

ANON. (1981) Action plan for dragonflies. *World Wildlife News* **11**, 2.

ANON. (1982) State Environment Protection Policy No. W-34B (The Waters of the Western District Lakes). *Victorian Government Gazette* No. 12, 11 February 1982, 431–52.

AUSTRALIAN CONSERVATION FOUNDATION (1972) *Pedder Papers. Anatomy of a Decision.* Australian Conservation Foundation, Melbourne.

AUSTRALIAN CONSERVATION FOUNDATION (1981) River and wetland modification. *Australian Conservation Foundation Policy No.* **28**, Appendix 4.

BOCKEL C. (1979) Notes on the status and behaviour of purple-crowned fairy-wren *Malurus coronatus* in the Victoria River Downs area, Northern Territory. *Aust. Bird Watcher* **8**, 91–7.

BURTON J.R., WILLIAMS W.D., ST. JOHN E. and HILL D.G. (1974) *The Flooding of Lake Pedder.* Lake Pedder Committee of Enquiry. Final Report, April 1974. Australian Government Publishing Service, Canberra.

COMMITTEE OF ENQUIRY INTO THE NATIONAL ESTATE (1974) *Report of the National Estate.* Australian Government Publishing Service, Canberra.

COSTIN A.B. and FRITH H.J. (eds) (1971) *Conservation.* Penguin Books, Ringwood.

COWAN I.M. (1973) The conservation of Australian waterfowl. *A.F.A.C. Special Publication* No. 2. Australian Government Publishing Service, Canberra.

COWLING S.J. (1979) The status of endangered waterfowl and wetlands in Australia. In M.J. Tyler (ed.) *The Status of Endangered Australasian Wildlife.* Royal Zoological Society of South Australia, Adelaide.

DUFFEY E. and WATT A.S. (eds) (1971) *The Scientific Management of Animal and Plant Communities for Conservation.* Blackwell Scientific Publications, Oxford.

EDWARDS R.W. and GARROD D.J. (eds) (1972) *Conservation and Productivity of Natural Waters.* Academic Press, London.

FENNER F. (1975) *A National System of Ecological Reserves in Australia.* Australian Academy of Science, Report No. 19, 1–114.

FRANKEL O.H. and SOULÉ M.E. (1981) *Conservation and Evolution.* Cambridge University Press, Cambridge.

FRITH H.J. (1973) *Wildlife Conservation.* Angus and Robertson, Sydney.

GOODRICK G.N. (1970) A survey of wetlands of coastal New South Wales. *CSIRO Div. Wildl. Res. Tech. Memo* **5**.

HOLLOWAY C. (1979) I.U.C.N., the Red Data Book, and some issues of concern to the identification and conservation of threatened species. In M.J. Tyler (ed.) *The Status of Endangered Australasian Wildlife.* Royal Zoological Society of South Australia, Adelaide.

IUCN (1980) *World Conservation Strategy.* IUCN, Gland, Switzerland.

JENKINS R.W.G. (1979) The status of endangered Australian reptiles. In M.J. Tyler (ed.) *The Status of Endangered Australasian Wildlife.* Royal Zoological Society of South Australia, Adelaide.

JONES W. (1978) *The Wetlands of the South-East of South Australia.* Nature Conservation Society of South Australia, Adelaide.

LAKE J.S. (1971) *Freshwater Fishes and Rivers of Australia.* Nelson, Sydney.

LAKE P.S. (1974) Conservation. In W.D. Williams (ed.) *Biogeography and Ecology in Tasmania.* Junk, The Hague.

LAKE P.S. (1974) Aquatic ecosystems — conflicting interests and conservation. *Aust. Soc. Limnol. Newsl.* **11**, 20–32.

LAKE P.S. (1978) On the conservation of rivers in Australia. *Aust. Soc. Limnol. Newsl.* **16**(2), 24–8.

LAKE P.S. (1980) Conservation. In W.D. Williams (ed.) *An Ecological Basis for Water Resource Management.* Australian National University Press, Canberra.

LAKE P.S. (1980) Littoral fauna of the Serpentine Impoundment. Paper given at

Annual Congress of Australian Society for Limnology, Queenstown, Tasmania, May 1980, pp. 1–15.

LEIGH J., BRIGGS J. and HARTLEY W. (1981) *Rare or Threatened Australian Plants.* Australian National Parks and Wildlife Service, Special Publication, **7**.

LUTHER H. and RZOSKA J. (1971) *Project Aqua: A Source Book of Inland Waters Proposed for Conservation.* Blackwell Scientific Publications, Oxford.

MARSHALL A.J. (ed.) (1966) *The Great Extermination.* Heinemann, London.

MITCHELL B.D. (in press) *Limnology of mound springs and temporary pools south and west of Lake Eyre.* Nature Conservation Society of South Australia, Adelaide.

MOIR W.H. (1972) Natural areas. *Science, N.Y.* **177**, 396–400.

MORGAN N.C. (1972) Problems of the conservation of freshwater ecosystems. *Symp. Zool. Soc. Lond.* No. 29, 135–54.

MORGAN N.C. (1978) Towards improved criteria for selection of wetlands for wildfowl conservation. In Proceedings, Technical Meeting for evaluation of wetlands from a conservation point of view, 15–19.

MYERS N. (1976) An expanded approach to the problem of disappearing species. *Science, N.Y.* **193**, 198–202.

OVINGTON D. (1978) *Australian Endangered Species. Mammals, Birds and Reptiles.* Cassell, Melbourne.

RABE F.W. and SAVAGE N.L. (1979) A methodology for the selection of aquatic natural areas. *Biol. Conserv.* **15**, 291–300.

RATCLIFFE D.A. (1976) Thoughts towards a philosophy of nature conservation. *Biol. Conserv.* **9**, 45–53.

RATCLIFFE D.A. (1977) Nature conservation: aims, methods and achievements. *Proc. Roy. Soc. Lond. B* **197**, 11–29.

RIGGERT T.L. (1966) *Wetlands of Western Australia: A Study of the Swan coastal plain.* Dept. Fish. Fauna, Perth.

SERVENTY V. (1966) *A Continent in Danger.* Andre Deutsch, London.

SHIEL R.J. (1980) Billabongs of the Murray-Darling system. In W.D. Williams (ed.) *An Ecological Basis for Water Resource Management.* Australian National University Press, Canberra.

SIMMONS I.G. (1974) *The Ecology of Natural Resources.* Edward Arnold, Melbourne.

SMITH A.J. (1975) A review of literature and other information on Victorian wetlands. *CSIRO Div. Land Use Res. Tech. Memo* **75/5**. [Similar reviews in the same publication series relate to New South Wales, South Australia, Western Australia, Tasmania].

SMITH L.A. and JOHNSTONE R.E. (1977) Status of the purple-crowned wren (*Malurus coronatus*) and buff-sided robin (*Poecilodryas superciliosa*) in Western Australia. *W. Aust. Nat.* **13**, 183–8.

SPECHT R.L. (1981) Conservation of vegetation types. In R.H. Groves (ed.) *Australian Vegetation.* Cambridge University Press, Cambridge.

SPECHT R.L., ROE E.M. and BOUGHTON V.H. (1974) Conservation of major plant communities in Australia and Papua New Guinea. *Aust. J. Bot. Suppl.* **7**, 1–667.

SPELLERBERG I.F. (1981) *Ecological Evaluation for Conservation.* Arnold, London.

STANTON J.P. (1975) A preliminary assessment of wetland areas in Queensland. *CSIRO Div. Land Use Res. Tech. Memo* **75/10**.

THE ROYAL SOCIETY OF LONDON (1977) *Scientific Aspects of Nature Conservation in Great Britain.* The Royal Society, London.

TIMMS B.V. (1977) Man's influence on dune lakes or the derogation of the delectable lakes of northeastern N.S.W. *Hunter Natural History,* August 1977, 132–41.

TURNER J.S. (1980) *Scientific Research in National Parks and Nature Reserves.* Australian Academy of Science, Canberra.

TYLER M.J. (1979) The impact of European man upon Australasian amphibians. In M.J. Tyler (ed.) *The Status of Endangered Australasian Wildlife.* Royal Zoological Society of South Australia, Adelaide.

TYLER P.A. (1976) Lagoons of Islands — death knell for a unique ecosystem? *Biol. Conserv.* **9**, 1–11.

UNESCO (1974) Criteria and guidelines for the choice and establishment of biosphere reserves. *Final Report. UNESCO Man and the Biosphere Programme* No. **22**.

WALKER K.F. (1982) The plight of the Murray crayfish in South Australia. *Red Gum* **6**(1), 2–6.

WATSON J.A.L. (1981) Odonata (dragonflies and damselflies). In A. Keast (ed.) *Ecological Biogeography of Australia.* Junk, The Hague.

WATTS C.H.S. (1979) The status of endangered Australian rodents. In M.J. Tyler (ed.) *The Status of Endangered Australasian Wildlife.* Royal Zoological Society of South Australia, Adelaide.

WEBB L.J., WHITELOCK D. and BRERETON J. LE GAY (eds) (1969) *The Last of Lands.* Jacaranda, Milton.

WELLS S.M. (Compiler) (1981) Extracts from: *IUCN Red Data Book for Invertebrates.* Conservation Monitoring Centre, Cambridge, U.K.

WILLIAMS W.D. (1981) The Crustacea of Australian inland waters. In A. Keast (ed.) *Ecological Biogeography of Australia.* Junk, The Hague.

ZWICK P. (1981) Blephariceridae. In A. Keast (ed.) *Ecological Biogeography of Australia.* Junk, The Hague.

12 THE IMPACT OF MAN

ANON. (1974) *Mine Waste Pollution of the Molonglo River. Final Report on Remedial Measures, June 1974.* Joint Government Technical committee on Mine Waste Pollution of the Molonglo River. Australian Government Publishing Service, Canberra.

ARTHINGTON A.H., CONRICK D.L., CONNELL, D.W. and OUTRIDGE, P.M. (1982) The ecology of a polluted urban creek. *AWRC Tech. Paper* No. 68.

BAYLY I.A.E. and LAKE P.S. (1979) *The Use of Organisms to Assess Pollution of Fresh Waters: A Literature Survey and Review.* Ministry for Conservation Environmental Studies Series, Pub. No. 258, Melbourne.

BLYTH J.D. (1980) Environmental impact of reservoir construction: the Dartmouth Dam invertebrate survey: a case history. In W.D. Williams (ed.) *An Ecological Basis for Water Resource Management.* Australian National University Press, Canberra.

BYCROFT B.M., COLLER B.A.W., DEACON G.B., COLEMAN D.J. and LAKE P.S. (1982) Mercury contamination of the Lerderberg River, Victoria, Australia, from an abandoned gold field. *Environ. Pollut. Ser. A.* **28**, 135–47.

CADWALLADER P.L. (1978) Some causes of the decline in range and abundance of native fish in the Murray-Darling River system. *Proc. Roy. Soc. Vict.* **90**, 211–24.

CAMPBELL I.C. (1978) A biological investigation of an organically polluted urban stream in Victoria. *Aust. J. Mar. Freshwat. Res.* **29**(3), 275–91.

CONNELL D.W. (1981) *Water Pollution. Causes and Effects in Australia and New Zealand.* 2nd edn. University of Queensland Press, Brisbane.

CROSBY R.G. (1973) The fate of pesticides in the environment. *Ann. Rev. Plant Physiol.* **24**, 467–92.

CULLEN P., ROSICH R.S. and BEK P. (1977) A phosphorus budget for Lake Burley Griffin and management implications for urban lakes. *AWRC Tech. Paper* No. 31.

DOOLAN K.J. and SMYTHE L.E. (1973) Cadmium content of some New South Wales waters. *Search* **4**(5), 162–3.

DUFFUS J.H. (1980) *Environmental Toxicology.* Edward Arnold, London.

FÖRSTNER U. and Wittman G.T.W. (1979) *Metal Pollution in the Aquatic Environment.* Springer-Verlag, Berlin.

FRASER J.C. (1972) Regulated discharge and the stream environment. In R.T. Oglesby, C.A. Carlson and J.A. McCann (eds) *River Ecology and Man.* Academic Press, New York.

HART B.T. (1974) A compilation of Australian water quality criteria. *AWRC Tech. Paper* No. 7.

HART B.T. and DAVIES S.H.R. (1978) A study of the physico-chemical forms of trace metals in natural waters and wastewaters. *AWRC Tech. Paper* No. 35.

HYNES H.B.N. (1960) *The Biology of Polluted Waters.* Liverpool University Press, Liverpool.

JEFFREE R.A. and WILLIAMS N.J. (1980) Mining pollution and the diet of the purple-striped gudgeon *Mogurnda mogurnda* Richardson (Eleotridae) in the Finniss River, Northern Territory, Australia. *Ecol. Monogr.* **50**(4), 457–85.

JERNELÖV A. and MARTIN A.-L. (1975) Ecological implications of metal metabolism by microorganisms. *Am. Rev. Microbiol.* **29**, 61–77.

KHAN M.A.Q. (ed.) (1977) *Pesticides in Aquatic Environments.* Plenum Press, New York.

KRENKEL P.A. and NOVOTNY V. (1980) *Water Quality Management.* Academic Press, New York.

LAKE P.S. (1979) Accumulation of cadmium in aquatic animals. *Chem. Aust.* **46**(1), 26–9.

LAKE P.S., SWAIN R. and MILLS B. (1979) Lethal and sublethal effects of cadmium on freshwater crustaceans. *AWRC Tech. Paper* No. 37.

LAWS E.A. (1981) *Aquatic Pollution.* Wiley, New York.

LOCKWOOD A.P.M. (ed.) (1976) *Effects of Pollutants on Aquatic Organisms.* Cambridge University Press, Cambridge.

MASON C.F. (1981) *Biology of Freshwater Pollution.* Longman, London.

MCIVOR C.C. (1976) The effects of organic and nutrient enrichments on the benthic macroinvertebrate community of Moggill Creek, Queensland. *Water* **3**(4), 16–21.

MELLANBY K. (1980) *The Biology of Pollution.* 2nd edn. Edward Arnold, London.

NAGY L.A. and OLSON B.H. (1980) Mercury in aquatic environments: A general review. *Water* **7**(3), 12–15.

NATIONAL ACADEMY OF SCIENCES, NATIONAL ACADEMY OF ENGINEERING (1973) *Water Quality Criteria 1972.* Environmental Protection Agency, Washington, D.C.

NORRIS R.H., SWAIN R. and LAKE P.S. (1981) Ecological effects of mine effluents on the South Esk River, north-eastern Tasmania. II. Trace metals. *Aust. J. Mar. Freshwat. Res.* **32**, 165–73.

RAVERA O. (ed.) (1979) *Biological Aspects of Freshwater Pollution.* Pergamon Press, Oxford.

SALÁNKI J. and BIRÓ P. (1979) *Human Impacts on Life in Fresh Waters.* Akadémiai Kiado, Budapest.

SENATE SELECT COMMITTEE (1970) *Water Pollution in Australia.* Commonwealth Government Printing Office, Canberra.

SHIEL R.J. (1978) Zooplankton communities of the Murray-Darling system, a preliminary report. *Proc. Roy. Soc. Vict.* **90**, 193–202.

SHIEL R.J. (1981) *Plankton of the Murray-Darling river system.* Ph.D. Thesis, University of Adelaide.

U.S. ENVIRONMENTAL PROTECTION AGENCY (1976) *Quality Criteria for Water.* U.S. Environmental Protection Agency, Washington, D.C.

VALLENTYNE J.R. (1972) Freshwater supplies and pollution: effects of the demophoric explosion on water and man. In N. Polunin (ed.) *The Environmental Future.* Macmillan, London.

VALLENTYNE J.R. (1974) *The Algal Bowl: Lakes and Man.* Department of the Environment, Fisheries and Marine Service, Ottawa.

VOLLENWEIDER R.A. (1968) *Scientific Fundamentals of the Eutrophication of Lakes and Flowing Waters, with Particular Reference to Nitrogen and Phosphorus as Factors in Eutrophication.* OECD, Paris.

WALKER K.F. (1980) The downstream influence of Lake Hume on the River Murray. In W.D. Williams (ed.) *An Ecological Basis for Water Resource Management.* Australian National University Press, Canberra.

WALKER K.F. (1982) Ecology of freshwater mussels in the River Murray. *AWRC Tech. Paper.* No. 63.

WALKER K.F., HILLMAN T.J. and WILLIAMS W.D. (1980) The effects of impoundments on rivers: an Australian case study. *Verh. Int. Ver. Limnol.* **20**, 1695–701.

WEATHERLEY A.H., BEEVERS J.R. and LAKE P.S. (1967) The ecology of a zinc-polluted river. In A.H. Weatherley (ed.) *Australian Inland Waters and their Fauna: Eleven Studies.* Australian National University Press, Canberra.

WEATHERLEY A.H., LAKE P.S. and ROGERS S.C. (1980) Zinc pollution and the ecology of the freshwater environment. In J. Nriagu (ed.) *Zinc in the Environment. Part 1: Ecological Cycling.* Wiley, New York.

WELCH E.B. (1980) *Ecological Effects of Waste Water.* Cambridge University Press, Cambridge.

WILLIAMS W.D. (1967) The changing limnological scene in Victoria. In A.H. Weatherley (ed.) *Australian Inland Waters and their Fauna: Eleven Studies.* Australian National University Press, Canberra.

WILLIAMS W.D. (1980) Catchment management. In W.D. Williams (ed.) *An Ecological Basis for Water Resource Management.* Australian National University Press, Canberra.

WILLIAMS W.D. (1980) Toxicity of wastewaters: Some general implications. *Water* **7**(3), 23–4.

WOOD G. (1975) An assessment of eutrophication in Australian inland waters. *AWRC Tech. Paper* No. 15.

13 RECORDS FROM THE PAST

BARNES M.A. and BARNES W.C. (1978) Organic compounds in lake sediments. In A. Lerman (ed.) *Lakes. Chemistry Geology Physics.* Springer-Verlag, New York.

BARTON C.E. (1978) *Magnetic studies of some Australian lakes.* Ph.D. Thesis, Australian National University, Canberra.

Bowler J.M. (1971) Pleistocene salinities and climatic change: Evidence from lakes and lunettes in south-eastern Australia. In D.J. Mulvaney and J. Golson (eds) *Aboriginal Man and Environment in Australia.* Australian National University Press, Canberra.

Bowler J.M. (1976) Aridity in Australia: age, origin and expression on aeolian landforms and sediments. *Earth Sci. Rev.* **12**, 279–310.

Bowler J.M. (1981) Australian salt lakes. A palaeohydrologic approach. In W.D. Williams (ed.) *Salt Lakes.* Junk, The Hague.

Bowler J.M. and Hamada T.T. (1971) Late Quaternary stratigraphy and radiocarbon chronology of water level fluctuations in Lake Keilambete, Victoria. *Nature, Lond.* **232**, 330–2.

Bowler J.M., Hope G.S., Jennings J.N., Singh G. and Walker D. (1976) Late Quaternary climates of Australia and New Guinea. *Quat. Res.* **6**, 359–94.

Churchill D., Galloway R.W. and Singh G. (1978) Closed lakes and their palaeoclimatic record. In A.B. Pittock, L.A. Frakes, D. Jensen, J.A. Peterson and W. Zillman (eds) *Climatic Change and Variability: A Southern Perspective.* Cambridge University Press, Cambridge.

Clark R.L. (1976) *Vegetation history and the influence of the sea at Lashmar's Lagoon, Kangaroo Island, South Australia, 3,000 B.P. to the present day.* B.Sc. Honours Thesis, Monash University, Melbourne.

Currey D.T. (1964) The former extent of Lake Corangamite. *Proc. Roy. Soc. Vict.* **77**, 377–86.

De Deckker P. (1981) *Taxonomy, ecology and palaeoecology of ostracods from Australian inland waters.* Ph.D. Thesis, University of Adelaide.

Dodson J.R. (1974) Calcium carbonate formation by *Enteromorpha nana* algae in a hypersaline volcanic crater lake. *Hydrobiologia* **44**, 247–57.

Dodson J.R. (1974) Vegetation history and water fluctuations at Lake Leake, south-eastern South Australia. I. 10,000 B.P. to present. *Aust. J. Bot.* **22**, 719–41.

Dodson J.R. (1974) Vegetation and climatic history near Lake Keilambete, Western Victoria. *Aust. J. Bot.* **22**, 709–17.

Dodson J.R. (1975) Vegetation History and water fluctuations at Lake Leake, south-eastern South Australia. II. 50,000 B.P. to 10,000 B.P. *Aust. J. Bot.* **23**, 815–31.

Dodson J.R. (1979) Late Pleistocene vegetation and environment near Lake Bullen Merri, western Victoria. *Aust. J. Ecol.* **4**, 419–27.

Dodson J.R. and Wilson I.B. (1975) Past and present vegetation of Marshes Swamp in south-eastern South Australia. *Aust. J. Bot.* **23**, 123–50.

Edmondson W.T. (1974) The sedimentary record of the eutrophication of Lake Washington. *Proc. Nat. Acad. Sci. USA* **71**, 5093–5.

Frey D.G. (1974) Palaeolimnology. *Mitt. Int. Ver. Limnol.* **20**, 95–123.

Galloway R.W. and Kemp E.M. (1980) Late Cainozoic environments in Australia. In A. Keast (ed.) *Ecological Biogeography in Australia.* Junk, The Hague.

Kershaw A.P. (1970) A pollen diagram from Lake Euramoo, north-east Queensland, Australia. *New Phytol.* **69**, 785–805.

Kershaw A.P. (1974) A long continuous pollen sequence from north-eastern Australia. *Nature, Lond.* **251**, 222–3.

Kershaw A.P. (1976) A late Pleistocene and Holocene pollen diagram from Lynch's Crater, north-eastern Queensland, Australia. *New Phytol.* **77**, 469–98.

Kershaw A.P. (1981) Quaternary vegetation and environments. In A. Keast (ed.) *Ecological Biogeography of Australia.* Junk, The Hague.

KRISHNASWAMI S. and LAL D. (1978) Radionuclide limnochronology. In A. Lerman (ed.) *Lakes. Chemistry Geology Physics.* Springer-Verlag, New York.

LEWIS W.M. JR and WEIBEZAHN F.H. (1981) Chemistry of a 7.5 m sediment core from Lake Valencia, Venezuela. *Limnol. Oceanogr.* **26**(5), 907–24.

LUDBROOK N.J. (1956) Microfossils from Pleistocene to Recent deposits, Lake Eyre, South Australia. *Trans. Roy. Soc. S. Aust.* **79**, 37–44.

MACKERETH F.J.H. (1966) Some chemical observations on post-glacial lake sediments. *Phil. Trans. Roy. Soc. B* **250**, 161–213.

MACPHAIL M.K. and JACKSON W.D. (1978) The late Pleistocene and Holocene history of the Midlands of Tasmania: pollen evidence from Lake Tiberias. *Proc. Roy. Soc. Vict.* **90**, 287–300.

PATERSON C.G. and WALKER K.F. (1974) Recent history of *Tanytarsus barbitarsis* Freeman (Diptera: Chironomidae) in the sediments of a shallow, saline lake. *Aust. J. Mar. Freshwat. Res.* **25**, 315–25.

PENNINGTON W. (Mrs. T.G. Tutin) (1979) The origin of pollen in lake sediments. *New Phytol.* **83**, 189–213.

PENNINGTON W. (Mrs. T.G. Tutin) (1981) Records of a lake's life in time: the sediments. *Hydrobiologia* **79**, 197–219.

POLACH H. and SINGH G. (1980) Contemporary ^{14}C levels and their significance to sedimentary history of Bega Swamp, New South Wales. *Radiocarbon* **22**, 398–409.

SINGH G. (1981) Late Quaternary pollen records and seasonal palaeoclimates of Lake Frome, South Australia. In W.D. Williams (ed.) *Salt Lakes.* Junk, The Hague.

SINGH G., OPDYKE N.J. and BOWLER J.M. (1981) Late Cainozoic stratigraphy, palaeomagnetic chronology and vegetational history from Lake George, Australia. *J. Geol. Soc. Aust.* **28**, 435–52.

TUTIN W. (1969) The usefulness of pollen analysis in interpretations of stratigraphic horizons, both Late-glacial and Post-glacial. *Mitt. Int. Ver. Limnol.* **17**, 154–64.

WALKER D. and SINGH G. (1981) Vegetation history. In R.H. Groves (ed.) *Australian Vegetation.* Cambridge University Press, Cambridge.

YEZDANI G.H. (1970) *A study of the Quaternary vegetation history in the volcanic lakes region of western Victoria.* Ph.D. Thesis, Monash University, Melbourne.

Index